Titles in This Series

Titles in This Series

Volume

38 **Topics in complex analysis,** Dorothy Browne Shaffer, Editor

39 **Errett Bishop: Reflections on him and his research,** Murray Rosenblatt, Editor

40 **Integral bases for affine Lie algebras and their universal enveloping algebras,** David Mitzman

41 **Particle systems, random media and large deviations,** Richard Durrett, Editor

42 **Classical real analysis,** Daniel Waterman, Editor

43 **Group actions on rings,** Susan Montgomery, Editor

44 **Combinatorial methods in topology and algebraic geometry,** John R. Harper and Richard Mandelbaum, Editors

45 **Finite groups—coming of age,** John McKay, Editor

46 **Structure of the standard modules for the affine Lie algebra $A_1^{(1)}$,** James Lepowsky and Mirko Primc

47 **Linear algebra and its role in systems theory,** Richard A. Brualdi, David H. Carlson, Biswa Nath Datta, Charles R. Johnson, and Robert J. Plemmons, Editors

48 **Analytic functions of one complex variable,** Chung-chun Yang and Chi-tai Chuang, Editors

49 **Complex differential geometry and nonlinear differential equations,** Yum-Tong Siu, Editor

50 **Random matrices and their applications,** Joel E. Cohen, Harry Kesten, and Charles M. Newman, Editors

51 **Nonlinear problems in geometry,** Dennis M. DeTurck, Editor

52 **Geometry of normed linear spaces,** R. G. Bartle, N. T. Peck, A. L. Peressini, and J. J. Uhl, Editors

53 **The Selberg trace formula and related topics,** Dennis A. Hejhal, Peter Sarnak, and Audrey Anne Terras, Editors

54 **Differential analysis and infinite dimensional spaces,** Kondagunta Sundaresan and Srinivasa Swaminathan, Editors

55 **Applications of algebraic K-theory to algebraic geometry and number theory,** Spencer J. Bloch, R. Keith Dennis, Eric M. Friedlander, and Michael R. Stein, Editors

56 **Multiparameter bifurcation theory,** Martin Golubitsky and John Guckenheimer, Editors

57 **Combinatorics and ordered sets,** Ivan Rival, Editor

58 **The Lefschetz centennial conference. Part I: Proceedings on algebraic geometry,** D. Sundararaman, Editor

58.II **The Lefschetz centennial conference. Part II: Proceedings on algebraic topology,** S. Gitler, Editor

58.III **The Lefschetz centennial conference. Part III: Proceedings on differential equations,** A. Verjovsky, Editor

59 **Function estimates,** J. S. Marron, Editor

60 **Nonstrictly hyperbolic conservation laws,** Barbara Lee Keyfitz and Herbert C. Kranzer, Editors

61 **Residues and traces of differential forms via Hochschild homology,** Joseph Lipman

62 **Operator algebras and mathematical physics,** Palle E. T. Jorgensen and Paul S. Muhly, Editors

63 **Integral geometry,** Robert L. Bryant, Victor Guillemin, Sigurdur Helgason, and R. O. Wells, Jr., Editors

64 **The Legacy of Sonya Kovalevskaya,** Linda Keen, Editor

65 **Logic and combinatorics,** Stephen G. Simpson, Editor

66 **Free group rings,** Narian Gupta, Editor

The Lefschetz Centennial Conference

Proceedings on Algebraic Topology

SOLOMON LEFSCHETZ

CONTEMPORARY MATHEMATICS

Volume 58.II

The Lefschetz Centennial Conference

Proceedings on Algebraic Topology

Proceedings of The Lefschetz Centennial Conference held December 10–14, 1984

S. Gitler, Editor

AMERICAN MATHEMATICAL SOCIETY

Providence · Rhode Island

The Proceedings of The Lefschetz Centennial Conference on Algebraic Geometry, Algebraic Topology and Differential Equations was held at the Centro de Investigación y de Estudios Avanzados, in Mexico City, Mexico, December 10-14, 1984.

1980 *Mathematics Subject Classification*. Primary 55-06.

Organizing committee: J. Adem, S. Gitler, J. J. Kohn, E. Ramirez de Arellano, D. Sundararaman, A. Verjovsky.

Editorial Committee of the Proceedings: S. Gitler, J. Adem, E. Ramirez de Arellano, D. Sundararaman, A. Verjovsky.

Library of Congress Cataloging-in-Publication Data
(Revised for part II)

Lefschetz Centennial Conference (1984: Mexico City, Mexico). The Lefschetz Centennial Conference.

(Contemporary mathematics, ISSN 0271-4132; v. 58)
Includes bibliographies.
Contents: pt. 1. Proceedings of algebraic geometry, D. Sundararaman, editor — pt. 2. Proceedings on algebraic topology / S. Gitler — pt. 3. Proceedings on differential equations / A. Verjovsky.
1. Algebraic geometry–Congresses. 2. Algebraic topology–Congresses. 3. Differential equations–Congresses. I. Lefschetz, Solomon, 1884-1972. II. Sundararaman, D. III. Gitler, Samuel. IV. Verjovsky A. V. Series: Contemporary mathematics (American Mathematical Society); v. 58.
QA564.L43 1984 512′.33 86-14040
ISBN 0-8218-5063-6 (pt. II)
ISBN 0-8218-5065-2 (set)

TABLE OF CONTENTS

PREFACE

During the week of December 10-14, 1984, the Department of Mathematics of the Centro de Investigación del IPN held an International Conference in Mexico City, to celebrate Solomon Lefschetz' 100th birthday.

This conference was made possible by the generous financial assistance of the Centro de Investigación del IPN, the Consejo Nacional de Ciencia y Tecnología, the Secretaría de Educación Pública (MEXICO) and the National Science Foundation (USA).

We thought it appropriate to have lectures in: Algebraic Geometry, Algebraic Topology and Differential Equations, the three main areas of Lefschetz' research. The Proceedings appear in three volumes and commemorate Lefschetz's contribution to the development of Mathematics in Mexico. This volume includes the papers on Algebraic Topology.

I want to thank the participants and especially the authors of this volume for making the conference a success.

Samuel Gitler.

CINVESTAV, Mexico City, November 1986.

Contemporary Mathematics
Volume 58, Part II, 1987

THE K-THEORY OF PROJECTIVE STIEFEL MANIFOLDS

Enrique Antoniano

Here I want to state some results I recently obtained in collaboration with Samuel Gitler and Jack Ucci (theorem 6), about the K-theory of projective Stiefel manifolds. I will begin by explaining what these manifolds are and the reason of our interest in all this.

If $k \leqslant n$, the Stiefel manifold $V_{n,k}$, consist of all orthonormal k-frames in R^n:

$$V_{n,k} = \{(v_1,\dots,v_k) \mid v_i \in R \text{ and } v_i \cdot v_j = \delta_{ij}\}$$

These manifolds are homogeneous spaces of the orthogonal group:

$$V_{n,k} = \{\text{left cosets of } O(n-k) \subset O(n)\} = O(n)/O(n-k)$$
$$A \mapsto \begin{pmatrix} A & 0 \\ 0 & I \end{pmatrix}$$

The projective Stiefel manifold $X_{n,k}$ is obtained by identifying each frame $(v_1,\dots,v_k)$, with its negative $(-v_1,\dots,-v_k)$ and it is also an homogeneous space of the orthogonal group:

$$X_{n,k} = O(n)/O(n-k) \times Z_2 \quad \text{where} \quad Z_2 = \{\pm I\} \subset O(n)$$

Now, we have a double covering $V_{n,k} \to X_{n,k}$ and this double covering is classified by a map:

$$f : X_{n,k} \to P^\infty$$

Consider the question suggested by the following diagram:

$$\begin{array}{ccc} & & X_{n,k} \\ & S? \nearrow & \downarrow f \\ P^m & \hookrightarrow & P^\infty \end{array}$$

Given n and k, what is the maximal value of m for which there exists a function s making the diagram commutative?

This question is related to other known problems as we can see from the next two propositions; see [1], [4]:

PROPOSITION 1: There is a map s making the preceeding diagram commutative if and only if there is a skew-linear and non singular map

$$R^{m+1} \times R^k \to R^n$$

PROPOSITION 2: There exists an immersion $P^{K-1} \looparrowright R^{n-1}$, if and only if there is a map s making the following diagram commutative:

$$\begin{array}{ccc} & & X_{n,k} \\ & \overset{s}{\nearrow} & \downarrow f \\ P^{K-1} & \hookrightarrow & P^{\infty} \end{array}$$

We would like to apply K-theory and Adams operations looking for obstructions to the existence of s in the next diagram:

$$\begin{array}{ccc} & & K(X_{n,k}) \\ & \overset{s^*}{\swarrow} & \uparrow f^* \\ K(P^m) & \xleftarrow{1^*} & K(P^{\infty}) \end{array}$$

Here, the values of $K(P^m)$, $K(P^{\infty})$, and i^* are well known and it is our objective to describe $K(X_{n,k})$ and f^*. With this purpose, we will make use of the following theorem due to Hodgkin, [6], [9].

Let $R(G)$ and $R(H)$ be the representation rings of G and H. Then, Z and $R(H)$ are $R(G)$ modules via the augmentation and the homomorphism induced by the inclusion of H in G.

THEOREM 3: Let G be a compact Lie group and $H \subset G$ a closed subgroup. Then there is a spectral sequence $\{E_r^{p,q}\}$ such that:

1) $$E_2^{p,q} = \begin{cases} \mathrm{Tor}^P_{R(G)}\ (R(H);\ Z) & \text{if } q \text{ is even} \\ 0 & \text{if } q \text{ is odd} \end{cases}$$

2) $\{E_r\}$ converges to $K(G/H)$

To apply such a theorem we will restrict ourwelves to the case $X_{4n,2k-1}$, since then:

$$X_{4n,2k-1} = \mathrm{Spin}(4n)/\mathrm{Spin}(4n-2k+1) \times Z_2$$

so the involved Lie groups have easier representation rings, in fact, [5], [8]:

$$R(\mathrm{Spin}(4n)) = Z[\pi_1,\dots,\pi_{2n-2}, X_{4n}, \delta^+_{4n}]$$

$$R(\mathrm{Spin}(4n-2k+1) \times Z_2) = Z[\pi_1,\dots,\pi_{2n-k-1}, \delta, y]/(y^2+2y)$$

To determine the algebra $\mathrm{Tor}_{R(G)}(R(H);\ Z)$, we have the following lemma that describes:

$$j^* : R(\mathrm{Spin}(4n)) \to R(\mathrm{Spin}(4n-2k+1) \times Z_2)$$

LEMMA 4: The homomorphism j^* is given by:

$$j^*(\pi_i) = (-1)^{i-1} \sum_{j=1}^{i} 2^{2j-1} \binom{2n-i+j}{j}\ y\ \pi_{i-j} + (1+y)^i \pi_i$$

$$j^*(X_{4n}) = -2^{k-1}\delta y - 2^{2n-1}\ y$$

$$j^*(\delta^+_{4n}) = 2^{k-1}\delta$$

where, in the right side of the first equation $\pi_i = 0$ if $i > 2n-k$ and

$$\pi_{2n-k} = \delta^2 + 2^{2n-k+1}\delta - \sum_{j=1}^{2n-k} 2^{2j}\pi_{2n-k-j}$$

PROOF. First, it is possible to give an explicit description for the restriction of j to the maximal toruses. Then, considering that the representation ring of a group is embedded in the one for its torus, one can use this description to obtain the lemma.

Now, we can follow A. Roux [9] and apply some results on homological algebra due to Cartan and Eilenberg [3], to obtain the next proposition.

PROPOSITION 5: The Hodgkin spectral sequence for $X_{4n,2k-1}$, collapses, and:

$$E_2^{*,q} = E_\infty^{*,q} = \begin{cases} E(\beta_1,\dots,\beta_{k-2},z_1,z_2,u)\otimes Z[y,\delta]/\sim & \text{if } q \text{ is even} \\ 0 & \text{if } q \text{ is odd} \end{cases}$$

where the exterior generators live in $E^{1,*}$ and the polynomial generators in $E^{0,*}$. Furthermore:

$$\sim \;=\; \begin{cases} 2^\alpha y = 0 \;;\; y^2 = -2y \\ 2^{k-1}\delta = 0 \;;\; \delta^2 = -2^{2n-k+1}\delta + by \\ 2^r u = 0 \\ z_1 y = z_1 u = 0 \\ z_2\delta = b/2^{\alpha-r}u \;;\; z_2 u = 0 \\ uy = -2u - 2^{k-1-r}\delta z_1 \\ u\delta = -2^{2n-k+1}u + 2^{\alpha-r}yz_2 \end{cases}$$

where

$$\alpha = \alpha(4n,2k-1) = \min\{2n-1, 2i-1+\upsilon_2\tbinom{2n}{i};\ i \geq 2n-k+1\}$$

$$r = \min\{\alpha, k-1\} \quad\text{and}\quad b = 2^{rn-2k+1}[1+(-1)^{k+1}\tbinom{2n-1}{2n-k}]$$

THEOREM 6: As a Z_2-graded algebra

$$K(X_{4n,2k-1}) = E(\beta_1,\dots,\beta_{k-2},z_1,z_2,u)\times Z[y,\delta]/\sim$$

where the exterior generators live in K^1 and the polynomial generators in K^0. Furthermore, the relations $\sim$ are described by the same formulas of proposition 5, except for the last two that now looks as follows:

$$uy = -2u - 2^{k-1-r}\delta z_1 + v_1$$

$$u\delta = -2^{2n-k+1}u + 2^{\alpha-r}yz_2 + v_2$$

where v_1 and v_2 are integer linear combinations of y and δ.

PROOF. As we can see from proposition 5, the algebra $E^{*,0}$, is a quotient of a universal algebra, i.e. a Z-graded algebra which is the tensor product of an exterior algebra with generators in dimension 1 by a polynomial algebra with generators in dimension 0. Also, the K-theory of a finite complex is a quotient of such a universal algebra when one considers only the natural Z_2-graduation.

Since the algebra K is an extension of the algebra E, it follows that both are quotients of the same universal algebra say by ideals I_K and I_E. Now we observe that in our case, the ideal I_E, is generated by elements in dimensions 0, 1 and 2, so this is also true for I_K. By inspection, we find the relations in theorem 6.

Let ξ denote the Hopf bundle over $X_{n,k}$, which is the line bundle associated with the double covering $V_{n,k} \to X_{n,k}$, and let $\rho \in K(X_{n,k})$ be the complexification of $(\xi-1) \in KO(X_{n,k})$.

COROLLARY 7. The order of $\rho \in K(X_{4n,s})$ is $2^{\alpha(4n,s)}$, where

$$\alpha(4n,s) = \min\{2n-1,\ 2i-1+\upsilon_2\tbinom{2n}{i};\ i \geq [\tfrac{4n-s+2}{2}]\}$$

PROOF. For $s = 2k-1$, we have the injection; [9]:

$$E_\infty^{0,0} \to K(X_{4n,2k-1})$$
$$y \mapsto \rho$$

so ρ has the same order as y.

On the other hand, the fibration $S^{4n-2k} \to X_{4n,2k} \to X_{4n,2k-1}$ is totally non cohomologous to zero in K-theory, [5], so ρ has the same order in $K(X_{4n,2k})$ as in $K(X_{4n,2k-1})$, as the corollary claims.

Let $2^{\beta(4n,s)}$ be the order of $(\xi-1) \in KO(X_{4n,s})$; then we have that $\beta(4n,2) = \alpha(4n,s)+\varepsilon$, with $\varepsilon = 0$ or 1. Now, if $n > 2$, $s < 4n-2$ and $(n,s) \neq (3,8)$ or $(4,8)$, it is possible to show that $4ns$ is not divisible by $2^{\beta(4n,s)}$ and since $4ns(\xi-1)$ is the stable tangent bundle for the manifold $X_{4n,s}$, [7], we conclude that it is not stably parallelizable.

Joining these facts to the explicit constructions of vector fields given by P. Zvengrowski in [10] and some others given later by himself [2], we get the following corollary.

COROLLARY 8. About the parallelizability of projective Stiefel manifolds, we have the next table:

Parallelizable	Not known	Not stably parallelizable
$X_{n,n}$ $X_{n,n-1}$ $X_{2n,2n-2}$ $X_{4,s}$ $X_{8,s}$ $X_{16,8}$	$X_{12,8}$	All not in a previous column

BIBLIOGRAPHY

1. J. Adem, S. Gitler & I.M. James. "On axial maps of a certain type". Bol. Soc. Mat. 17 (1972), 59-62.

2. E. Antoniano, S. Gitler, J. Ucci, P. Zvengrowski. "The projective Stiefel manifolds, K-theory and parallelizability" (to appear).

3. H. Cartan and S. Eilemberg. "Homological Algebra". Princeton University Press, Princeton Math. series, 19 (1965).

4. S. Gitler. "The projective Stiefel manifolds II. Applications". Topology 7 (1968), 47-53.

5. S. Gitler & K.Y. Lam. "The K-theory of Stiefel manifolds". Springer Verlag. Lecture Notes in Math. 168 (1970), 35-66.

6. L.H. Hodgkin and V.P. Snaith. "Topics in K-theory, Two independent contributions". Springer-Verlag. Lecture Notes in Mathematics 496 (1975).

7. K.Y. Lam. "Formula for the tangent bundle of flag manifolds and related manifolds". Trans. Math. Soc. 213 (1975), 305-311.

8. J. Milnor. "The representation rings of some classical groups". Notes for Math. 402 (1963).

9. A. Roux. "Application de la suite espectrale d'Hodgkin au calcul de la K-theorie des varietes de Stiefel". Bull. Soc. Mat. France 99 (1971), 345-468.

10. P. Zvengrowski. "Ueber die Parallelisierbarkeit von Stiefel Mannigfaltigkeit". Forschunginstitut für Mathematik ETH Zurich und University of Calgary. April, 1976.

DEPARTMENT OF MATHEMATICS
CENTRO DE INVESTIGACION Y DE ESTUDIOS AVANZADOS-IPN
Apdo. Postal 14-740
07000 Mexico 14, D.F. MEXICO

Contemporary Mathematics
Volume **58**, Part II, 1987

BP OBSTRUCTION THEORY

Martin Bendersky

ABSTRACT. A secondary BP obstruction theory for desuspending a truncated projective space is developed. Computations are made which are sufficient to show that RP^{57} does not **embed** in R^{100}.

1. INTRODUCTION. A generalized homology theory E_* may be used to obtain obstructions to desuspending a C.W. complex. Briefly if E is a reasonable spectrum, i.e. a ring spectrum with unit, and X is a finite C.W. complex, we define $E(X)$ to be $\varinjlim_k \Omega^k(E_k \wedge X)$ and define a map $\eta: X \longrightarrow E(X)$ by adjointing the Hurewicz homomorphism (see [5]). If in addition E satisfies the flatness condition (Adams [1]) we may define a coaction

$$\psi: E_*(X) \longrightarrow E_*(E) \otimes_{E_*} E_*(X)$$

the map ψ must factor through $E_*(\eta)$. If X is an odd desuspension of a truncated, real, projective space and E is the BP-spectrum, then, through a range, the factorization of ψ forces $BP_*(X)$ to be an unstable comodule in the sense of [5; 1.1]. This in turn implied very strong non-desuspensions for truncated real projective spaces.

There is more information contained in this factorization. The stablization map $BP_*(BP(X)) \rightarrow BP_*(BP) \otimes_{BP_*} BP_*(X)$ has a kernel which was not exploited in [5]. In this paper we indicate how this kernel may be used to obtain stronger non-desuspensions.

Anyone familiar with [5] is aware of the computational complexity inherent in unstable BP obstruction theory. As might be expected the secondary theory is a bit more unpleasant. With this in mind we shall only sketch the method to a point where we can prove the following non-embedding.

1980 Mathematics Subject Classification: Primary 55N22, 55P40; secondary 57R42.

THEOREM 1.1. $RP^{57} \not\subset R^{100}$.

This had improved previous results by two dimensions and is our first new result. Astey [3] has since proven the following.

THEOREM (Astey). $RP^{2(m+\alpha(m)-1)} \not\subset R^{4m-2\alpha(m)+1}$.

In particular Astey implies $RP^{57} \not\subset R^{99}$. Since our improvement is only one dimension, we only include a sketch of the method which may be of independent interest for the problem of maximally desuspending a truncated projective space.

We thank Don Davis for many valuable conversations. Details and further applications will appear elsewhere.

We use the following notation:

P_a^b is the truncated, real projective space RP^b/RP^{a-1}.

$\alpha(n)$ is the number of ones in the binary expansion of n.

$\nu(n)$ is the exponent of 2 in the primary decomposition of n.

2. UNSTABLE STRUCTURES. Recall $BP_* = \pi_* BP$ is a polynomial algebra over $Z_{(2)}$ (the integers localized at 2) on generators $u_i \in BP_{2(2^i-1)}$. The generators of Araki [2] are the most convenient for unstable computations.

Let $h_i \in BP_{2(2^i-1)}(BP)$ be the element $c(t_i)$ where c is the canonical antiautomorphism and t_i are the elements defined in [1]. Then $BP_* BP \simeq BP_*[h_1, h_2, \cdots]$. We shall denote $BP_* BP$ by Γ_*. Γ_* is a right and left BP_* module. The relation between the two module structures is given by the integral form of Ravenel's formula [5;2.1].

For Y a space $V(Y)$ denotes the functor described in [4]. We first recall V for a free, graded BP_*-module M.

DEFINITION 2.1. $V(M) = BP_*\text{-span}\{h^I \otimes_{BP_*} m \mid 2\ell(I) \leqslant |m|\} \subset \Gamma_* \otimes_{BP_*} M$ where $I = (i_1, i_2, \cdots)$, $h^I = h_1^{i_1} h_2^{i_2} \cdots$ and $\ell(I) = i_1 + i_2 + \cdots$.

For a general, graded BP_*-module choose an exact sequence $F_1 \xrightarrow{f} F_0 \longrightarrow M \longrightarrow 0$ with F_i free.

DEFINITION 2.2. (a) $V(M) = \operatorname{coker} V(f)$ and (b) $V(Y) = V(BP_*(Y))$.

We now restrict Y to be a desuspension of a truncated, real projective space. The following computation of $BP_*(P_a^b)$ is due to I. Hansen and S. Wilson.

THEOREM 2.3. $BP_*(P_{2n+1}^{2m-\varepsilon})$, $\varepsilon = 0,1$ is the BP_*-module generated by elements $\beta_i \in BP_{2i+1}(P_{2n+1}^{2m-\varepsilon})$, $n \leqslant i < m$ with relations r_i , $n \leqslant i < m - \varepsilon$ where $r_i = \sum_{j=0}^{i-n} c_j \beta_{i-j}$ with $c_j \in BP_{2j}$ defined by the 2-series (see [5]) $[2]\, T = \sum_{i \geqslant 0} c_j T^{j+1}$. The coaction is given by

$$\psi(\beta_i) = \sum_{j=1}^{i} \Big(\sum_{k \geqslant 0}^{F^*} h_k \Big)_{i-j}^{j} \otimes \beta_j$$

where $\sum^{F^*} = c \sum^{F} c$, $\sum^{F}$ the formal sum for BP and c the anti-isomorphism of Γ_*. $BP_*(P_{2n}^{b}) \simeq BP_*(P_{2n-1}^{b}) \oplus BP_*(S^{2n})$ as Γ_*-comodules.

It is convenient to tensor $V(Y)$ with $BP_*/(u_3, u_4, \cdots)$ and work modulo h_i , $i \geqslant 3$. We may choose a basis for the relations in theorem 2.3 in which the appearance of u_1 is restricted to only one term.

THEOREM 2.4. (Davis [6]) The relations are generated by

$$2\beta_i = u_1 \beta_{i-1} + u_2 \beta_{i-3} + \sum \alpha_j u^J \beta_{i-1/2|u^J|}$$

where $\alpha_j \in Z_{(2)}$ and $J = (0, j_2, j_3, \cdots)$

Let $Y = \Sigma^{-(2t+1)} P_{2w+1}^{2w+2d+\varepsilon}$, $\varepsilon = 0,1$ with $4(w-t) > 2d + 2\varepsilon - 6$.

Let F_0 be a free BP_*-module on generators β_i' corresponding to the generators of $BP_*(Y)$, and F_1 a free BP_*-module with generators r_i' corresponding to the relations in 2.4. There is a short exact sequence

$$(2.5) \qquad 0 \longrightarrow F_1 \xrightarrow{\ f\ } F_0 \longrightarrow BP_*(Y) \longrightarrow 0 \ .$$

Let $F_t' = \bigvee_k S^{n_{k,t}}$, $t = 0,1$, $n_{k,t}$'s corresponding to β_i''s or r_j''s .

In [5] a fibration

$$(2.6) \qquad BP(F_0') \longrightarrow BP(Y) \xrightarrow{\theta} BP(\Sigma F_1')$$

of ∞-loop maps was constructed. The map θ classified the fibration $BP(F_1') \xrightarrow{f'} BP(F_0') \longrightarrow BP(Y)$ with $\pi_*(f') = f$. In [5] we proved that the coaction ψ must lift to $V(F_0)$. Using the BP-Serre spectral sequence of the fibration 2.6, we may prove the following.

THEOREM 2.7. There is a commutative diagram

$$\begin{array}{ccc} & V(Y) & \\ {}^{\psi_U}\nearrow & & \searrow^{\sigma} \\ BP_*(Y) & \xrightarrow[\psi]{\qquad\qquad} & \Gamma_* \otimes_{BP_*} BP_*(Y) \end{array} .$$

In other words the relations in $BP_*(Y)$ must be compatible with the relations in $V(Y)$.

DEFINITION 2.8. $K = K_Y = \ker\{V(Y) \xrightarrow{\sigma} \Gamma_* \otimes_{BP_*} BP_*(Y)\}$.

As we shall see, the obstructions that arise from 2.7 naturally lie in K.

3. THE STRUCTURE OF K.

Let $\gamma_i \in BP_*(Y)$ denote the desuspension of β_{w+i}. We shall use the unstable indexing for elements in $V(Y)$, i.e. $h_2^d h^a u^I \gamma_i$ denotes the element $h_2^d h^{a+b} \otimes u^I \gamma_i$ where b is the dimension of γ_0 and $h = h_1$. We define the <u>degree</u> of $h_2^d h^a u^I \gamma_i$ to be $3d + a + 1/2|u^I| + i$. Its dimension is $2(3d + a + i + 2b) + |u^I|$. To prove 2.1 we need knowledge of $V(Y)$ through degree 6.

For $x = c\, h_2^d h^a u^I \gamma_t$ the excess, $e(x)$ was defined in [5] to be $d + a - t - 1/2|u^I| - \nu_2(c)$. The relation $u_1 = 1 \cdot u_1 - 2h$ allows us to lift elements of excess $\leqslant 0$ to $V(Y)$.

PROPOSITION 3.1. In degrees $\leqslant 6$ $V(Y)$ is generated over $Z_{(2)}$ by:

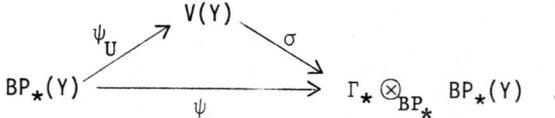

$u^j h_2^t h^i \gamma_a$ $\quad t + i \leqslant a$

$u^j h_2^t h^i u_r \gamma_a$ $\quad a \leqslant i + t \leqslant a + 2^r - 1$

$2^k u^j h_2^t h^i u_r^\varepsilon \gamma_a$ $\quad k > 0$, excess = 0 and $\varepsilon = 0$ or 1

where $u = u_2$ and $r \neq 1$.

Sketch of Proof. We start with a good set of generators for $V(Y)$. By Wilson [7;1.6] $V(Y)$ is generated over BP_* through degree 6 by

(a) $h_2^t h^i \gamma_a$, $t + i \leqslant a$;

(b) $h_2^t h^i u_1 \gamma_a$, $a \leqslant i + t \leqslant a + 1$; and

(c) $h_2^t h^i u_k \gamma_a$, $a \leqslant i + t \leqslant a + 2^k - 1$, $k > 1$.

The relations in 2.4 allow us to write generators of type (b) in terms of (a) and (c). u_1 acting on the left may inductively be transformed into a linear combination generators in the theorem by using the formula $u_1 = 1 \cdot u_1 - 2h$.

Elements in K arise in two ways. First, there are the obvious terms of the form $2^{t+1} h^a \gamma_t$ of excess $\leqslant 0$, e.g. $2h\gamma_0$. More subtle terms are a consequence of the right action formula for u_2 . The right action on u_2 defines polynomials $u^n = q_n(h, h_2, u_1, u)$ where u_1 and u $(= u_2)$ are acting on the right. We may replace u_1 with $-2S + uS^{-2}$ where S is the shift operator satisfying $S^j\gamma_i = \gamma_{i+j}$. (Here and in the sequel, we are ignoring units and only using the first terms of 2.4 in this replacement.) We have:

$$(3.2) \quad q_1 = 2^2 h^3 + h^2 u S^{-2} - 2h^2 S + hu^2 S^{-4} + 2^2 h S^2 - 2^2 h u S^{-1} + u + 2h_2$$

and modulo terms with excess $\leqslant 0$

$$(3.3) \quad q_1 = 2^2 h^3 + h^2 u S^{-2}$$

$$q_2 = 2^4 h^6 + h^4 u^2 S^{-4} + 2^3 h^5 u S^{-2} .$$

(The excess of $c h^j u^i S^t$ is $j - 3i - \nu_2(c) - t$.)

Example. We have $u h^0 \gamma_0 = 2^2 h^3 \gamma_0$ plus terms in $V(Y)$. (This follows from 3.3.) $2^2 h^3 \gamma_0$ is not in $V(Y)$ but must be a linear combination of generators from 3.1. Using 3.2 we see that $u h^0 \gamma_0 + 2 h^2 \gamma_1 - 2^2 h \gamma_2 - 2 h^0 h_2 \gamma_0 - h^0 u \gamma_0$ stabilizes to $2^2 h^3 \gamma_0$ and therefore represents a non-zero element in K. In a similar way, $u h \gamma_1 = 2^2 h^4 \gamma_1$, $u^2 h^0 \gamma_0 = 2^4 h^6 \gamma_0$ and $u(2h_2 \gamma_0) = 2^3 h^3 h_2 \gamma_0$ modulo excess $\leqslant 0$ determine elements of K.

DEFINITION 3.4.

$$k_1 = u h^0 \gamma_0 + 2 h^2 \gamma_1 - 2^2 h \gamma_2 - h^0 u \gamma_0 - 2 h^0 h_2 \gamma_0 ,$$

$$k_2 = u h \gamma_1 - 2 h^3 \gamma_2 - 2^2 h^2 \gamma_3 - h u \gamma_1 + 2 h h_2 \gamma_1 ,$$

$$k_3 = u^2 h^0 \gamma_0 + 2^2 h^4 \gamma_2 + 2^4 h^2 \gamma_4 + 2^4 h^5 \gamma_1 + h^0 u^2 \gamma_0 + 2^2 h_2^2 \gamma_0 + 2^4 h^3 h_2 \gamma_0 + 2^3 h^2 h_2 \gamma_1 ,$$

$$k_4 = u(h^{-1} h_2 \gamma_0) + 2 h h_2 \gamma_1 + 2^2 h_2 \gamma_2 + h^{-1} h_2 u \gamma_0 + 2 h^{-1} h_2^2 \gamma_0 .$$

Remark. The term $h^0 u^2 \gamma_0$ in k_3 is equal to a sum of generators from 3.1. However, we will not need this for our computation.

We obtain relations in K from the action of q_i on relations in $BP_*(Y)$ or by multiplying stable relations by powers of 2 to bring the excess down to zero. An example of the first type of relation is given by $u(2 h^0 \gamma_0) = 2^2 h^2 \gamma_1 - 2^3 h^3 \gamma_0$. An example of the second type of relation is given by $2k_1 = 2(2^2 h^3 \gamma_0)$. Since $2k_1 = 2^2 h^2 \gamma_1$, we obtain the same relation. By systematically examining all possible cases in 3.1, we can prove the following.

THEOREM 3.5. Through degree 6, K has the following BP_* generators: $2^{i+1} h_2^t h^j \gamma_i$ with $t + j - 2i - 1 \leqslant 0$; k_1 ; k_2 ; $2 h k_2$; k_3 ; k_4. Through degree 6, the relations are generated by: $(2h)^\ell (2^a h^a \gamma_a) = (2h)^t (2^b h^b \gamma_b)$ if $\ell + 2a = t + 2b$; $2^{i+1} h^i \gamma_0 = 0$; $2 k_1 = 2^2 h^2 \gamma_1$; $2(2h)^\varepsilon k_2 = (2h)^\varepsilon 2^3 h^4 \gamma_1$, $(\varepsilon = 0,1)$; $2 k_3 = 2^3 h^4 \gamma_2$; $u(2^3 h^3 \gamma_0) = 2^4 h^5 \gamma_1 - 2^3 h^4 \gamma_2 + 2^3 h^2 h_2 \gamma_1$; $2^4 h^3 h_2 \gamma_0 = 2^3 h^2 h_2 \gamma_1$.

4. SKETCH OF PROOF OF 1.1. An <u>immersion</u> $RP^{2n-1} \subseteq R^m$ implies a desuspension of a truncated real projective space. Assuming $RP^{57} \subseteq R^{100}$, we have $gd((2^{L+1} - 58)\xi_{57}) \leqslant 43$. Hence there must be a desuspension of $P^{2^{L+1}-1}_{2^{L+1}-58}$ to a complex with bottom cell in dimension 43. If we have an embedding, then the Thom-Pontrjagin map implies the existence of a primitive with leading term γ_{28} (the top cell). By mapping into $BP_*(T(\nu \oplus \xi))$ (ν = the normal bundle of the embedding). We introduce a relation on the top cell and we obtain a primitive of the form $\gamma_{28} + q\,u\,\gamma_{26} + u^2(x)$, $q \in Z_{(2)}$. We claim q is odd. To see this we compute the stable coproduct mod 2, $(2,u_1,u_2,\cdots)^2$ and (h^2). If q were even we would have $0 = (\gamma_{28}) = h\,u_1 \otimes \gamma_{26} \neq 0$. (We are using the fact that the formal group law coefficient $a_{1,3}$ is zero mod 2.)

The top relation in $BP_*(T(\nu \oplus \xi))$ has the form $Re = 2\gamma_{28} - u_1\gamma_{27} + u\gamma_{25} + \sum \alpha_j u^j \gamma_{28-3j}$ modulo $(u_3,u_4,\cdots)$. The following lemma is the main computational result.

LEMMA 4.1. $0 = \psi_U(Re) = 2^3 h^4 \gamma_2 + (2K + u_1K + uK)$.

<u>Proof</u>. The strategy is to move all u's to the right of h's in $\psi(Re)$. We find that $\psi_U(Re) = 2^3 h^4 \gamma_2 + \sum x_i \bmod (2K + u_1K + uK)$ where $\bar{e}(x_i) \leqslant 0$. (For $x = c\,h^I u^J \gamma_k$, $c \in Z_{(2)}$, $\bar{e}(x)$ is defined to be $|I| - 1/2|u^J| - k$, see [5].) We must have $\sum x_i = 0$ stably. Since $\bar{e}(x_i) \leqslant 0$, $\sum x_i$ must be zero in V. The term $2^3 h^4 \gamma_2$ comes about in the following way: $u_1\psi(\gamma_{27}) = u_1(1+h)^{2^L-27}_{25} \otimes \gamma_2$ + other terms $= u_1(2^2 h^{25} \otimes \gamma_2)$ + o.t. (stable indexing) $= u_1(2^2 h^3 \gamma_2)$ + o.t. (unstable indexing) $= 2^3 h^4 \gamma_2$ + o.t.. 1.1 now follows since by inspection, $2^3 h^4 \gamma_2$ does not lie in $2K + u_1K + uK$ consistent with the relation $\gamma_{28} + u\gamma_{25} + \cdots = 0$ and the coassociativity of ψ_U. For example we have $2^3 h^4 \gamma_2 = u_1(u(2^2 h^2 \gamma_0) + 2^3 h^4 \gamma_1 + 2^2 h_2 h \gamma_1)$. Hence the unstable parts of the coaction on γ_{25} and γ_{28} are zero and $\psi_U(\gamma_{27})$ has the form $(u(2^2 h^2 \gamma_0) + 2^3 h^4 \gamma_1 + 2^2 h_2 h \gamma_1) + h \otimes \gamma_{26} + (2^2 h u_1 + 2h^2) \otimes \gamma_{25} + h^2 u_1 \otimes \gamma_{24} + \cdots$. Taking coproducts again we have in $\Gamma \otimes V(Y)$
$0 = 2^2 h^2 \otimes h^2 \gamma_1 + h \otimes \psi_U \gamma_{26} + h^2 u_1 \otimes \psi_U \gamma_{24}$. (We are using the stable indexing in the left factor and the unstable indexing in the right factor. We are also using the fact that the stable terms cancel among themselves.) The

relation $2\gamma_{25} = u_1\gamma_{24} + u\gamma_{22} + \cdots$ implies $u_1\gamma_{24}$ contributes no unstable terms. Since $2^2 h^2 \otimes h^2 \gamma_1$ does not cancel we have a contradiction. All other possible relations are delt with in the same way.

Remark. It is easy to give a complete description of the BP-Thom-Pontrjagin primitive τ. Let $\beta \in BP_2(CP^\infty)$ be the canonical class. Then, using the H-structure of CP^∞ we have a primitive β^k. Let $(\beta^k)_n$ be the image of β^k in $BP_*(CP_n^\infty)$ for $L \gg 0$, $x = (\beta^{2^L-1})_{2^L-n}/(2^L-1)!$ is integral. Then $\tau = f_*(x)$ where $f\colon \Sigma CP^{2^L-1}_{2^L-n} \longrightarrow P^{2^{L+1}}_{2^{L+1}-2n}$ is the canonical map. β^k is recursively computed in [4, pg. 748].

BIBLIOGRAPHY

[1] Adams, J.F., Stable Homotopy and Generalized Homology, University of Chicago Press, 1974.

[2] Araki, S., Typical formal groups in complex cobordism and K-theory, Kinokuniya, Tokyo, 1973.

[3] Astey, L., "A cobordism Obstruction to Embedding Manifolds", to appear.

[4] Bendersky, M., "Some Computations in the Unstable Adams-Novikov Spectral Sequence", Publ. Res. Inst. Math. 16 (1980), 739-760.

[5] ______, Davis, D.M., "Unstable BP-Homology and Desuspensions", to appear in Amer. J. Math.

[6] Davis, D.M., "A Strong Non-Immersion Theorem for Real Projective Space", Annals of Math. 120 (1984), 517-528.

[7] Wilson, W.S., "Brown Peterson Metastability and the Bendersky-Davis Conjecture", Publ. Res. Inst. Math. 20 (1984), 1037-1051.

DEPARTMENT OF MATHEMATICS
RIDER COLLEGE
LAWRENCEVILLE, NEW JERSEY 08648

Contemporary Mathematics
Volume **58**, Part II, 1987

CONTINUOUS COHOMOLOGY

Edgar H. Brown, Jr. and Robert H. Szczarba

ABSTRACT. This talk reports on work in progress aimed at carrying out an analogue of the Quillen-Sullivan rational homotopy theory in the case of real homotopy theory and continuous cohomology. The main result announced is a generalization of Van Est's theorem to simplicial Lie Groups and, as an application, the computation of the continuous cohomology of $K(R,n)$.

1. INTRODUCTION. In this talk I will be reporting on some joint work in progress with Robert Szczarba. Broadly speaking, we have been attempting to extend the machinery of homotopy theory so as to be able to better deal with $B\Gamma_q$, the classifying space for foliations. From a homotopy point of view $B\Gamma_q$ has a number of bizarre features. for example, for many values of i, $\pi_i(B\Gamma_q)$ maps onto the real numbers; also, $H^p(BO_q j;G) \to H^p(B\Gamma_q;G)$ is an injection for G the integers and zero for G the reals and $p > 2q$. Following standard procedures one can use a Postnikov tower to construct examples of spaces with these unusual properties, but it seems unlikely that such constructions could properly represent the homotopy type of $B\Gamma_q$. The work of Bott and Haefliger ([1]) suggests that one should be incorporating continuous cohomology into homotopy theory to deal with these situations.

The following may be a simple example of phenomena involved in $B\Gamma_q$. If G is an abelian topological group, let $K(G,n)$ denote the simplicial group given by the classical model for $K(G,n)$, namely, $K(G,n)_q = Z^n(\Delta_q;G)$ the group of normalized cocycles on the simplicial set $\Delta_q = \{(i_0,i_1,\ldots,i_p) | 0 \leqq i_0 \leqq \cdots \leqq i_p \leqq q\}$ with coefficients in G. We topologize $K(G,n)_q$ by viewing it as a finite cartesian product of copies of G. Thus $K(G,n)$ is a simplicial abelian topological group. We realize it as a space in the usual way, namely,

$$|K(G,n)| = \cup \; |\Delta_q| \times K(G,n)_q/\sim$$

We let G^d denote G with the discrete topology. Since $|K(G,n)|$ is an abelian topological group, its Postnikov system is trivial and one may check that

---

1980 Mathematics Subject Classification. 55B99,53D20

$\pi_{i+n}(|K(G,n)|) \approx \pi_i(G)$. Thus for $G = R/Z = S^1$,

$$|K(R/Z,n)| \simeq |K(Z,n+1)|$$

LEMMA 1.1. If $i: K((R/Z)^d,n) \to K(R/Z,n)$ is the identity map, then $|i|: |K((R/Z)^d,n)| \to |K(R/Z,n)|$ corresponds to the Bockstein operation δ^* associated to $0 \to Z \to R \to R/Z \to 0$.

Suppose $u \in H^n(BO_q;Z)$ is an element of infinite order, for example, a power of $\mathbf{P}_1$; represent u by a map

$$u: BO_q \to |K(R/Z,n-1)| = |K(Z,n)|$$

Let $v = u + |i|$ and $\bar{v} = u + id$ in the diagram

$$\begin{array}{ccc} BO_q \times |K(R/Z)^d,n-1)| & \xrightarrow{u+|i|} & |K(R/Z,n-1)| \\ \downarrow id \times |i| & \nearrow & \\ BO_q \times |K(R/Z,n-1)| & u + id & \end{array}$$

Let E be the fibration induced by v from the contractible fibration and let p be the composition

$$p: E \to BO_q \times |K(R/Z)^d,n-1)| \to BO_q$$

This is a fibration over BO_q in which p^*u is non-zero but p^*u is in the image of δ^* and hence zero as a class with real coefficients, (in forming E we killed $u + |i| = u + \delta^* L_{n-1}$.)

The fibre of p is the fibration over $|K((R/Z)^d,n-1)|$ induced by $|i|$, that is, $|K(R^d,n)|$. Using $\bar{v}$ instead of v we obtain a fibration $\bar{p}: \bar{E} \to BO_q$ with fibre $|K(R,n)|$ which is contractible via the homotopy coming from $R \times I \to R$ by $(x,t) \to tx$. Note that the underlying sets of E and $\bar{E}$ are identical and $E \to \bar{E}$ is the identity map. The pair $(E,\bar{E})$ is an example of a space with two topologies as dealt with in ([2]); if one defines $\pi_i(E,\bar{E})$ to be $\pi_i(E)$ with the topology induced from $\bar{E}$, $\pi_n(E,\bar{E}) \approx R$ as topological groups.

The remainder of this talk is devoted to the homotopy theory of spaces with two topologies. Actually, we find it more convenient to use simplicial spaces. A simplicial space X is a (semi) simplicial set X such that X_q has a topology for each q and the face and degeneracy operators are continuous. If G is a topological abelian group, $H^q(X;G)$ is the cohomology groups based on the cochains:

$$C^q(X;G) = \text{continuous maps of } X_q \text{ to } G$$

(The cohomology of X as a simplicial set is denoted by $H^q(X^d;G)$.)
If $(Y,\bar{Y})$ is a space with two topologies as above, it corresponds to the simplicial space $\Delta(Y,\bar{Y})$ consisting of the singular simplices of Y topologized as a subspace of $\Delta(\bar{Y})$, with the compact open topology. conversely, one passes from simplicial spaces to pairs by X goes to $(|X^d|,|X|)$.

At present we are working on the details of carrying over the Quillen-Sullivan rational homotopy to real homotopy, replacing simplicial sets by simplicial spaces and cohomology with rational coefficients with continuous cohomology with real coefficients (Hereafter $H^*(X) = H^*(X;R)$). This project is proving surprisingly difficult because a number of very elementary things do not carry over. For example, the fact that chain groups are free abelian does not carry over to the continuous case. In consequence we have not discovered how to use the Serre spectral sequence in this setting. We have completed a first step in this project, namely:

THEOREM 1.2. For any $n \geq 1$, $H^*(K(R,n)$ is a free graded commutative algebra on a single generator $\iota_n \in H^n(X)$.

Thus, $H^*(X)$ is a polynomial algebra on ι_n if n is even and an exterior algebra on ι_n if n is odd.

The main ingredient of the proof of Theorem 1.2 is a generalization of the Van Est Theorem which we now describe.

2. VAN EST'S THEOREM. Let $G = \{G_q\}$ be a simplicial Lie group. In analogy with the classical case, we define the continuous cohomology of G, $H^*(G)$, to be the homology of the cochain complex $(C^*(G),d)$ where

$$C^q(G) = \{f:G_q^q \to R: f \text{ continuous}\},$$

$G_q^q = G_q \times \ldots \times G_q$, q factors, and

$$(df)(g_1,\ldots,g_{q+1}) = f(\partial_0 g_2,\ldots,\partial_0 g_{q+1})$$
$$+ \sum_{i=1}^{q} (-1)^i f(\partial_i g_1,\ldots,\partial_i g_i \cdot \partial_i g_{i+1},\ldots,\partial_i g_{q+1})$$
$$+ (-1)^{q+1} f(\partial_{q+1} g_1,\ldots,\partial_{q+1} g_q).$$

If each $G_q = G$, a fixed Lie group, and if all face and degeneracy mappings are the identity, then this definition agrees with the usual definition of the continuous cohomology of G. (See Van Est [3].)

The Lie algebra $L(G)$ of the simplicial Lie group G is the simplicial Lie algebra with $L(G)_q = L(G_q)$, the usual Lie algebra of the Lie group G_q. The face and degeneracy mappings for $L(G)$ are the differentials of those for G. For H a simplicial Le subgroup of G, we define a double complex$(C^*(L(G(,H)d_1,d_2)$ by letting $C^{p,q}(L(G),H)$ be the space of linear mappings

$$\alpha:\Lambda^p (G_q) \to R$$

with $\alpha(v_1 \wedge \ldots \wedge v_p) = 0$ whenever some $v_i \in (H_q)$ and with

$$\alpha(Ad(h)v_1 \wedge\ldots\wedge (Ad(h)v_p) = \alpha(v_1 \wedge\ldots\wedge v_p)$$

for $h \in H_q$, $v_i \in L(G_q)$. (Here $\Lambda^p L(G_q)$ is the pth exterior power of $L(G_q)$.) The differentials

$$d_1:C^{p,q}(L(G),H) \to C^{p+1,q}(L(G),H)$$
$$d_2:C^{p,q}(L(G),H) \to C^{p,q+1}(L(G),H)$$

are given by

$$(d_1\alpha)(v_1 \ldots v_{p+1}) = \sum_{0\le i<j\le p+1} (-1)^{i+j}\alpha([v_i,v_j]\ v_1 \ldots \hat{v}_i \ldots \hat{v}_j \ldots v_{p+1})$$

$$(d_1\alpha)(w_1 \ldots w_p) = \sum_{i=0}^{q+1} (-1)^i\ \alpha(\partial_i w_1 \ldots \partial_i w_p)$$

for $\alpha \in C^{p,q}(L(G),H), v_i \in L(G_q), w_j \in L(G_{q+1})$. The Lie algebra cohomology of $L(G)$ rel H is defined to be the homology of the total complex associated to the double complex $(C^*(L(G),H),d_1,d_2)$.

Using the usual formulas one sees that both the continuous cohomology of G and the Lie algebra cohomology of $L(G)$ relative to H are graded commutative algebras.

A simplicial Lie subgroup K of G is maximal compact if each K_q is a maximal compact subgroup of G_q. Our version of the Van Est Theorem is the following.

THEOREM 2.1. Let K be a maximal compact subgroup of the simplicial Lie group G. Then

$$H^*(G) \approx H^*(L(G),K)$$

as graded algebras.

The proof of this result is a variant on the original proof of Van Est (see Van Est [3]).

3. $H^*(K(R,n))$. Recall, $K(R,n)_q$ is a finite dimensional real vector space and hence a Lie group. Let $G(n)$ be $K(R,n)$ viewed as a simplicial Lie group. We continue to view $K(R,n)$ as a simplicial space. A standard argument shows

LEMMA 3.1. $H^*(G(n)) \approx H^*(K(R,n+1))$

We now apply theorem 2.2 to $G(n)$. Since $G(n)_q = R^{m_q}$, its maximal compact subgroup is zero and its Lie brackets are zero. Hence $L(G(n)) = G(n)$ and in the double complex $C^{p,q} = C^{p,q}(L(G(n)),0)$, the differential $d_1 = 0$. Let Λ denote the function which assigns to a vector space V is exterior algebra $\Lambda(V)$. Viewing $G(n)$ as a simplicial vector space, $\Lambda(G(n))$ is a simplicial vector space and

$$H^*(L(G(n))) = \pi_*\ \Lambda(G(n)))^*$$

where $(\ \ )^* = \mathrm{Hom}(\ \ ,R)$.

Let $\overline{\Lambda}$ be the function which assigns to a graded vector space U its exterior algebra $\overline{\Lambda}\,(U)$, in the graded sense:

$$a \wedge b = -(-1)^{pq}\ b \wedge a,\ a \in U_p,\ b \in U_{q^*}$$

Then standard FD arguments show that $\pi_*(\Lambda^*(G(n))) \approx \overline{\Lambda}(\pi_*(G(n)))$. Thus

$$H^*(K(R,n+1) \approx \overline{\Lambda}(\pi_*(G(n)))^*$$

and $\pi_i(G(n)) = R\ \ i=n$ and $=0$ otherwise. The additive part of Theorem 1.2 now follows. The ring structure is determined by similar methods.

BIBLIOGRAPHY

1. Bott, R. and Haefliger, A., On the characteristic classes of Γ-foliations, Bull. Amer. Soc., 78(1972), 1039-1044.

2. Mostow, M.A., Continuous Chohomology of Spaces With Two Topologies, Memoirs of A.M.S., Providence, 1948.

3. Van Est, Group cohomology and Lie algebra, cohomology in Lie groups I, II, Nederl. Akad. Wetensch. Proc. Ser. A, 56(1953), 484-492, 493-504.

DEPARTMENT OF MATHEMATICS
BRANDEIS UNIVERSITY
WALTHAM, MASSACHUSETTS, 02254

DEPARTMENT OF MATHEMATICS
YALE UNIVERSITY
NEW HAVEN, CONNECTICUT 06520

Contemporary Mathematics
Volume 58, Part II, 1987

HOMOLOGY OF MAPPING CLASS GROUPS FOR SURFACES OF LOW GENUS

F.R. Cohen[1]

ABSTRACT. Let $\Gamma^n_{g,k}$ denote the mapping class group of a connected orientable surface $M^n_{g,k}$ of genus g with k boundary components and n marked points. Thus $\Gamma^n_{g,k}$ is $\pi_0\, Top^+(M^n_{g,k})$ where Top^+ denotes the group of orientation preserving homeomorphisms. W. Magnus gave a presentation of $\Gamma^n_{0,0}$ in 1934 [10]. J. Birman gave a presentation of $\Gamma^0_{2,0}$ [3, 4]. Using these presentations, we obtain some direct computations for $H_*(\Gamma^n_{0,0};R)$ and $H_*(\Gamma^0_{2,0};R)$. Work of R. Lee and S. Weintraub [9], of which we had been unaware until this conference, gave analogous results for $H_*(\Gamma^0_{2,0};R)$ if 30 is a unit in R. Similar computations have been done by R.J. Milgram. Finally, there are analogues of the Dyer-Lashof construction for certain $\Gamma^n_{g,k}$ and these are used to show that a certain classical map from Artin's braid group to the mapping class group is usually trivial in homology.

We thank the Centro de Investigacion for their hospitality during this conference and W. Singer for his suggestions.

1. PRESENTATIONS

Let Γ^n denote $\Gamma^n_{0,0}$ throughout the rest of this note. W. Magnus gave the following presentation for Γ^n [10]:

(i) There are generators $\sigma_1,\ldots,\sigma_{n-1}$ (given by Dehn twists about adjacent punctures).

(ii) A complete set of defining relations is given by

(a) $\sigma_i\sigma_j = \sigma_j\sigma_i$ if $|i-j| \geq 2$,

(b) $\sigma_i\sigma_{i+1}\sigma_i = \sigma_{i+1}\sigma_i\sigma_{i+1}$ for all i, and

(c) $(\sigma_1\sigma_2\cdots\sigma_{n-1})^n = 1 = (\sigma_1\sigma_2\cdots\sigma_{n-1})(\sigma_{n-1}\cdots\sigma_2\sigma_1)$.

---

[1]Partially supported by the NSF.

A presentation for Artin's braid group on n strands, B_n, is given by (i) together with relations (a) and (b) in (ii) [2]. A presentation for the symmetric group of n letters, Σ_n, is given by the above presentation for B_n together with the relation $\sigma_i^2 = 1$ for all i.

There is a natural quotient map

$$\gamma : B_n \to \Gamma^n.$$

Magnus computed the kernel of γ and showed that it is isomosphic to $Z \times F_{n-1}$ where F_{n-1} is a free group on $(n-1)$ letters. Thus there is a commutative diagram of group extensions.

$$\begin{array}{ccccc}
Z \times F_{n-1} & \xrightarrow{i} & P_n & \xrightarrow{\alpha} & K_n \\
\downarrow & & \downarrow & & \downarrow \\
Z \times F_{n-1} & \xrightarrow{j} & B_n & \xrightarrow{\gamma} & \Gamma^n \\
\downarrow & & \downarrow \pi & & \downarrow \lambda \\
1 & \longrightarrow & \Sigma_n & \xrightarrow{=} & \Sigma_n
\end{array}$$

where P_n and K_n are the kernels of π and λ respectively. P_n is the pure braid group on n strands [13, 2, 3].

2. IMPLICATIONS

Proposition 2.1: $H^*(K_n;\mathbb{Z})$ (with trivial action) is finitely generated and torsion free with Poincaré series $(1+2t)(1+3t)\ldots(1+(n-2)t)$. Furthermore $H^*(K_n;\mathbb{Z})$ is generated as an algebra by $H^1(K_n;\mathbb{Z})$ and $H^*(K_n;\mathbb{Z})$ is a sub Σ_n-module of $H^*(P_n;\mathbb{Z})$.

We remark that $H^*(P_n;\mathbb{Z})$ as an algebra over Σ_n has been given in [1] and [7, pp. 259-270]. Let p be a prime and $\mathbb{F}_p$ the field with p elements. The groups $H^*(\Sigma_n;H^*(P_n;\mathbb{F}_p))$ have been studied in [7, p. 279] where it is observed that $H^0(\Sigma_n;H^*(P_n;\mathbb{F}_p))$ vanishes in degrees other than 0 or 1. We shall check that the composite

$$H^0(\Sigma_n;H^*(P_n;\mathbb{F}_p)) \longrightarrow H^*(P_n;\mathbb{F}_p) \longrightarrow H^*(Z \times F_{n-1}, \mathbb{F}_p)$$

is a monomorphism if $p > n$. Furthermore, since Birman and Hilden [3, 4] have exhibited a central extension

$$\mathbb{Z}/2 \longrightarrow \Gamma^0_{2,0} \longrightarrow \Gamma^6_{0,0},$$

the following corollary of 2.1 follows at once.

Corollary 2.2: $\bar{H}^*(\Gamma^n;\mathbb{Z})$ (with trivial action) is all p-torsion for primes with $p \le n$. $\bar{H}^*(\Gamma^0_{2,0};\mathbb{Z})$ is all 2, 3 and 5 torsion.

The result in 2.2 for $\Gamma^0_{2,0}$ was given in [10].

By identifying the embedding of $H^*(K_n;\mathbb{Z})$ given by 2.1, one can do more precise computations. For example, one has

<u>Corollary 2.3</u>: There is a homomorphism $f\colon\ \mathbb{Z}/p \longrightarrow \Gamma^{p+1}$ such that f is an isomorphism in mod-p homology for all primes p. Thus $\Sigma B\mathbb{Z}/p \longrightarrow \Sigma B\Gamma^{p+1}$ is a p-local equivalence.

Recall that the 2-Sylow subgroup of Σ_4 is $\Sigma_2 \int \Sigma_2$, the dihedral group of order 8. Let $M(2)$ denote the cofibre of the natural map $B\Sigma_2 \int \Sigma_2 \longrightarrow B\Sigma_4$. As is well known by work of Mitchell and Priddy [11], $M(2)$ is the homotopy direct limit of the self-map e_2 of $\Sigma(B\Sigma_2 \times B\Sigma_2)$ given by

$$1 + \begin{pmatrix} 0 & 1 \\ 1 & 0 \end{pmatrix} + \begin{pmatrix} 1 & 1 \\ 0 & 1 \end{pmatrix} + \begin{pmatrix} 1 & 0 \\ 1 & 1 \end{pmatrix}$$

which is the second Steinberg idempotent in $\mathbb{Z}_{(2)}[GL_2(\mathbb{F}_2)]$.

<u>Proposition 2.4</u>: The mod-2 cohomology of Γ^4 is isomorphic to that of $B\Sigma_4 \vee M(2)$. The mod-3 cohomology of Γ^4 is that of $\mathbb{Z}/3$ (by Prop. 3.3). The classifying space of $B\Gamma^4$ is stably equivalent to

$$B\Sigma_4 \vee M(2) \vee Z$$

where Z is the stable summand of $B\mathbb{Z}/3$ with non-trivial H_1.

Mitchell and Priddy have determined the stable homotopy type of $M(2)$ [11] in terms of symmetric products.

Although we do not yet have a complete computation of $H^*(\Gamma^0_{2,0})$, we record the following observation.

<u>Proposition 2.5</u>. The map $\gamma\colon\ B_n \longrightarrow \Gamma^n$ factors through B_n/Z_n where Z_n is the center of B_n. Furthermore there is a lift

$$\begin{array}{ccc} & & \Gamma^0_{2,0} \\ & \nearrow & \downarrow \\ B_6/Z_6 & \longrightarrow & \Gamma^6_{0,0}, \end{array}$$

the induced map $H^*(\Gamma^{n,0}_0;\mathbb{F}_2) \longrightarrow H^*(B_n/Z_n;\mathbb{F}_2)$ is an epimorphism if $n \equiv 2(4)$, and $H^*(B_n/Z_n;\mathbb{F}_p)$ is isomorphic to $H^*(B_n;\mathbb{F}_p) \otimes \mathbb{F}_p[u]$ for u of degree 2 if p divides $n(n-1)$.

There are homomorphisms $D : B_{2q} \to \Gamma_{q,1}$ and $\Theta : B_p \int \Gamma_{q,1} \to \Gamma_{pq,1}$ given in section 5 here. Thus there are Dyer-Lashof operations defined in the homology of these groups (which are presumably well-known).

<u>Lemma 2.6.</u> The map D_* commutes with the Dyer-Lashof operations.

<u>Corollary 2.7</u>. If $g \geq 3$, D_* is trivial in mod-p homology.

Clearly Γ^n acts on $H_*(S^2-\{n \text{ points}\})$. Thus there is a natural representation $h\colon \Gamma^n \longrightarrow GL_{n-1}(\mathbb{Z})$ given by the action of Γ^n on $H_1(S^2-\{n \text{ points}\}; \mathbb{Z})$.

Proposition 2.8. The stable map $M(2) \longrightarrow B\Gamma^4$ given in 2.4 composed with Bh is stably null. Thus the composite

$$H_*(M(2);\mathbb{F}_2) \longrightarrow H_*(B\Gamma^4;\mathbb{F}_2) \longrightarrow H_*(GL_3(\mathbb{Z});\mathbb{F}_2)$$

is trivial.

3. PROOF OF 2.1

We first give the proof of 2.1. Consider the short exact sequence

$$\mathbb{Z} \times F_{n-1} \xrightarrow{j} B_n \longrightarrow \Gamma^n .$$

Magnus gave a presentation for $\mathbb{Z} \times F_{n-1}$ in terms of σ_i [10]. It is generated by

$$\begin{aligned} z &= (\sigma_1 \cdots \sigma_{n-1})^n \\ \tau_1 &= (\sigma_1 \cdots \sigma_{n-2})^{n-1} \\ \tau_2 &= (\sigma_1 \cdots \sigma_{n-3})^{n-2}\sigma_{n-1}^2 \\ \tau_q &= (\sigma_1 \cdots \sigma_{n-q-1})^{n-q}(\sigma_{n-q+1} \cdots \sigma_{n-1})^q \\ &\ \vdots \\ \tau_{n-1} &= (\sigma_2 \cdots \sigma_{n-1})^{-n+1} . \end{aligned}$$

Thus by direct calculation we have

Lemma 3.1. The structure of $H^*(\mathbb{Z} \times F_{n-1})$ as a Γ^n-module factors through the quotient map $\Gamma^n \longrightarrow \Sigma_n$. Hence the Hochschild-Lyndon-Serre spectral sequence for $\mathbb{Z} \times F_{n-1} \xrightarrow{i} P_n \xrightarrow{\alpha} K_n$ has trivial local coefficients.

Next, define

$$\bar{A}_{i+1,j} = \sigma_j\sigma_{j+1} \cdots \sigma_{i-1}\sigma_i\sigma_{i-1}^{-1} \cdots \sigma_{j+1}^{-1}\sigma_j^{-1} \quad \text{for } i \geq j.$$

It was shown in [7, pp. 254-270] that the Hurewicz homomorphism

$$\phi\colon P_n \longrightarrow H_1(P_n;\mathbb{Z})$$

carries $\bar{A}_{i+1,j}$ to a basis A_{i+1,j^*} for $n \geq i+1 > j \geq 1$. Furthermore, in the dual basis, the elements $A_{i+1,j}$ generate H^*P_n as an algebra with relations $A_{ij}^2 = 0$, and $A_{ij}A_{ik} = A_{kj}(A_{ik} - A_{ij})$, $j < k$. We remark that these last relations are a translation of the antisymmetry relation and Jacobi identity in a graded Lie algebra; this point will be elaborated elsewhere. The action of $\sigma \in \Sigma_n$ is given by

$$\sigma A_{ij} = A_{\sigma i,\sigma j} \quad \text{with } A_{ji} = A_{ij} \text{ if } j < i .$$

Since

$$(\sigma_1 \cdots \sigma_{q-1})^q = (A_{q,1}A_{q-1,1}\cdots A_{2,1})(A_{q,2}A_{q-1,2}\cdots A_{32})\cdots(A_{q,q-2}\,A_{q-1,q-2})A_{q,q-1}$$

it follows that

$$i_*(z) = \sum_{n\geq i>j\geq 1} A_{ij*}, \text{ and}$$
$$i_*(\tau_L) = \sum_{q-L\geq i>j\geq 1} A_{ij*} - \sum_{L\geq i>j\geq 1} A_{q-L+i,q-L+j}\,*.$$

Thus i^* is onto by inspection. Hence 2.1 follows directly from the collapsing spectral sequence with $E_2 = H^*(K_n;\mathbb{Z}) \otimes H^*(\mathbb{Z}\times F_{n-1};\mathbb{Z})$, and the fact that the Poincaré series for H_*P_n is $(1+t)(1+2t)\ldots(1+(n-1)t)$.

4. PROOFS OF 2.2 TO 2.5.

To prove 2.2, it suffices to check that the module of invariants $(\overline{H}^*(K_n;\mathbb{F}))^{\Sigma_n}$ is trivial for $\mathbb{F}=\mathbb{Q}$ or $\mathbb{F}=\mathbb{Z}/p$, $p>n$. But $(\overline{H}^*(P_n;\mathbb{F}))^{\Sigma_n}$ is concentrated in degree one with basis $A = \sum_{n\geq i>j\geq 1} A_{ij}$ [7, p. 270]. Dually, the formulas for i^* give that $i^*(A)\neq 0$ if $\mathbb{F}=\mathbb{F}_p$, $p>n$, or $\mathbb{F}=\mathbb{Q}$. Indeed $i^*(A) = nz^* +$ others.

The proof of 2.3 is analogous to that of 2.4. We give the proof of 2.4. By the computation of i^* we see that $H^1(K_4;\mathbb{Z}) = \mathbb{Z}\oplus\mathbb{Z}$ with generators

$$\alpha = A_{32} - A_{42} - A_{31} + A_{41} \quad \text{and}$$
$$\beta = A_{42} - A_{43} - A_{21} + A_{31}\,.$$

The action of Σ_4 on $H^1(K_4;\mathbb{Z})$ is thus given by the following chart where $(i,i+1)$ is the transposition acting as stated.

	(12)	(23)	(34)
α	$-\alpha$	$\alpha+\beta$	$-\alpha$
β	$\alpha+\beta$	$-\beta$	$\alpha+\beta$

We use this to compute $H^*(K_4;\mathbb{F}_p)$. Notice that

$$H^*(\mathbb{Z}/3\,;H^*(K_4;\mathbb{F}_3)) \cong \mathbb{F}_3 \quad \text{for all } q$$

by direct computation. It thus follows that

$$H^p(\Sigma_4\,;H^q(K_4;\mathbb{F}_3)) = \begin{cases} \mathbb{F}_3 & \text{if } q=0,\ p\equiv 0,3(4) \\ \mathbb{F}_3 & \text{if } q=1,\ p\equiv 0,1(4) \\ 0 & \text{otherwise} \end{cases}$$

because the normalizer of $\mathbb{Z}_3$ in Σ_4 acts on the spectral sequence. Thus $E_2 = E_\infty$ because $\mathbb{Z}/3$ is a retract of $\Gamma_0^{4,0}$ and E_2 in total degree n is $\mathbb{F}_3$.

To compute $H^*(\Sigma_4\,;H^*(K_4;\mathbb{F}_2))$ notice that we may regrade $H^*(K_4;\mathbb{F}_2)$ so that it is concentrated in degree 0. By inspection, this Σ_4-module is isomorphic to the Σ_4-module $\mathbb{F}_2[\Sigma_4/\Sigma_2\int\Sigma_2]$. Thus $H^*(\Sigma_4;\mathbb{F}_2[\Sigma_4/\Sigma_2\int\Sigma_2])$ is just $H^*(\Sigma_2\int\Sigma_2;\mathbb{F}_2)$. Thus

$$H^*(\Sigma_4; H^*(K_4; \mathbb{F}_2)) = H^*(\Sigma_4; \mathbb{F}_2) \oplus \Sigma\,(\text{Coker})$$

where Σ (Coker) is the module cokernel $(H^*(\Sigma_4; \mathbb{F}_2) \to H^*(\Sigma_2 \int \Sigma_2; \mathbb{F}_2))$ with homological degree increased by one. Next observe that there is a lift g

$$\begin{array}{ccc} & & \Gamma^4 \\ & \overset{g}{\nearrow} & \downarrow \lambda \\ \Sigma_2 \int \Sigma_2 & \subset & \Sigma_4 . \end{array}$$

Hence the Hochshild-Lyndon-Serre spectral sequence for λ collapses because the map $H^*(\Sigma_4; \mathbb{F}_2) \longrightarrow H^*(\Sigma_2 \int \Sigma_2; \mathbb{F}_2)$ is monic.

To obtain the stated stable decomposition, it suffices to exhibit a group homomorphism $\rho: \mathbb{Z} \times \Sigma_2 \times \Sigma_2 \longrightarrow \Gamma^4$ such that the composite

$$H_*(M(2); \mathbb{F}_2) \longrightarrow H_*(B\mathbb{Z} \times B\Sigma_2 \times B\Sigma_2; \mathbb{F}_2) \longrightarrow (H_*\Gamma^4; \mathbb{F}_2) \xrightarrow{\text{project}} H_*(\Sigma_4; H_1(K_4, \mathbb{F}_2))$$

is an isomorphism. Let G denote the subgroup of Σ_4 generated by $x = \sigma_1\sigma_2^2\sigma_1^{-1}$, $y = \sigma_1\sigma_3^{-1}$, and $z = \sigma_1\sigma_2\sigma_1\sigma_2^{-1}\sigma_3^{-1}\sigma_2^{-1}$. By Magnus' presentation one sees immediately that G is abelian with $y^2 = z^2 = 1$. Notice that the subgroup M generated by x factors through P_4; a generator of $H_1(M;\mathbb{Z})$ has image A_{31*} in integer homology. Thus G is isomorphic to $\mathbb{Z} \times \Sigma_2 \times \Sigma_2$ and there is a commutative diagram

$$\begin{array}{ccc} \mathbb{Z} & \longrightarrow & K_4 \\ \downarrow & & \downarrow \\ \mathbb{Z} \times \Sigma_2 \times \Sigma_2 & \longrightarrow & \Gamma^4 \\ \downarrow & & \downarrow \\ \Sigma_2 \times \Sigma_2 & \longrightarrow & \Sigma_4 . \end{array}$$

Next, let W denote a free $\mathbb{F}_2[\Sigma_4]$ resolution of F_2. By the above diagram we obtain a map $H_*(\Sigma_2 \times \Sigma_2; H_1(\mathbb{Z}; \mathbb{F}_2)) \longrightarrow H_*(\Sigma_4; H_1(K_4; \mathbb{F}_2))$. Using our identification $H_1(K_4; \mathbb{F}_2) \oplus \mathbb{F}_2 \cong \mathbb{F}_2(\Sigma_4/\Sigma_2 \int \Sigma_2)$ and the fact that the map $M \to K_4$ sends the generator of $H_1(\mathbb{Z}; \mathbb{F}_2)$ to $\alpha_* + \beta_*$ in $H_1(K_4; \mathbb{F}_2)$, we determine this homological map on the chain level: The composite θ

$$W \otimes_{\Sigma_2^2} H_1(\mathbb{Z}, \mathbb{F}_2) \longrightarrow W \otimes_{\Sigma_4} H_1(K_4; \mathbb{F}_2) \longrightarrow W \otimes_{\Sigma_4} \mathbb{F}_2[\Sigma_4/_{\Sigma_2} \textstyle\int \Sigma_2]$$

is given by $W \otimes 1 \longrightarrow W \otimes [(e + \sigma) + (\sigma + \sigma^2)]$ where σ^i denotes the coset $\sigma^i(\Sigma_2 \int \Sigma_2)$ in Σ_4 for $\sigma^i = (123)^i$. Since $W \otimes_{\Sigma_4} \mathbb{F}_2[\Sigma_4/\Sigma_2 \int \Sigma_2]$ isomorphic to $W \otimes_{\Sigma_2 \int \Sigma_2} \mathbb{F}_2$, we may transfer back to $W \otimes_{\Sigma_2^2} \mathbb{F}_2 \cong W \otimes_{\Sigma_2^2} H_1(M; \mathbb{F}_2)$. The composite (transfer $\cdot\, \theta$) is given by

$$(\text{transfer} \cdot \theta)(W \otimes 1) = W \otimes 1 + W(132) \otimes 1 + W(12) \otimes 1 + W(23) \otimes 1$$

by direct computation where $(132) = \sigma^2$ and $(i, i+1)$ denotes the evident transposition in Σ_4. Identifying $GL_2(\mathbb{F}_2)$ with Σ_3, one sees that this last composite is given by the Steinberg idempotent e_2. The splitting follows.

Next notice that Γ^4 acts on $H_*(S^2 - \{4 \text{ points}\})$. Thus there is a representation $h\colon \Gamma^4_{0,0} \longrightarrow GL_3(\mathbb{Z})$ given by the induced action on $H_1(S^2 - \{4 \text{ points}\};\mathbb{Z})$. Clearly

$$h(\sigma_1) = \begin{pmatrix} 0 & 1 & 0 \\ 1 & 0 & 0 \\ 0 & 0 & 1 \end{pmatrix}, \quad h(\sigma_2) = \begin{pmatrix} 1 & 0 & 0 \\ 0 & 0 & 1 \\ 0 & 1 & 0 \end{pmatrix}, \text{ and}$$

$$h(\sigma_3) = \begin{pmatrix} 1 & 0 & 1 \\ 0 & 1 & 1 \\ 0 & 0 & 1 \end{pmatrix}.$$ Since $h(\sigma_1\sigma_2^2\sigma_1^{-1}) = 1$, the homomorphism $\rho\colon \mathbb{Z} \times \Sigma_2 \times \Sigma_2 \longrightarrow \Gamma^4$ composed with h factors through $\Sigma_2 \times \Sigma_2$. It follows from the remarks in the above paragraph that $M(2) \longrightarrow B\mathbb{Z} \times B\Sigma_2 \times B\Sigma_2 \longrightarrow BGL_3(\mathbb{Z})$ is stably null-homotopic. Proposition 2.8 follows.

We now prove 2.5. First notice that the center of B_n, Z_n, is isomorphic to $\mathbb{Z}$ with generator $(\sigma_1 \ldots \sigma_{n-1})^n$ by Chow [5]. The abelianization map (Hurewicz homomorphism) $\phi\colon B_n \to \mathbb{Z}$ induces a morphism of short exact sequences

$$\begin{array}{ccccc} Z_n & \xrightarrow{\ i\ } & B_n & \longrightarrow & B_n/Z_n \\ \downarrow\cong & & \downarrow\phi & & \downarrow \\ \mathbb{Z} & \xrightarrow{n(n-1)} & \mathbb{Z} & \longrightarrow & \mathbb{Z}/_{n(n-1)} \end{array}$$

where ϕ is an isomorphism on H^1. Thus i^* is trivial on $H^1(\ ;\mathbb{F}_p)$ if p divides $n(n-1)$. Thus $H^*(B_n/Z_n;\mathbb{F}_2)$ is isomorphic to $H^*(B_n;\mathbb{F}_2) \otimes H^*(\mathbb{C}P^\infty;\mathbb{F}_2)$ as was computed in [6]. We remark that if $p \nmid n(n-1)$, then $H^*(B_n/\mathbb{Z}_n;\mathbb{F}_p)$ is also given in [6].

Since there is a commutative diagram

$$\begin{array}{ccc} B_n & \xrightarrow{\ \gamma\ } & \Gamma^n \\ & \searrow^{\pi} & \downarrow\lambda \\ & & \Sigma_n, \end{array}$$

and $H^*(\Sigma_n;\mathbb{F}_2) \to H^*(B_n;\mathbb{F}_2)$ is an epimorphism, it follows that γ^* is an epimorphism on $H^*(\ ;\mathbb{F}_2)$.

Next compute $H_1(\Gamma^n;\mathbb{Z})$. By the presentation given by Magnus, it follows that $H_1(\Gamma^n;\mathbb{Z})$ is $\mathbb{Z}/g$ where $g = g.c.d.(n(n-1), 2(n-1))$. Thus

$$H_1(\Gamma^n;\mathbb{Z}) \cong \begin{cases} \mathbb{Z}/_{n-1} & \text{if } n \equiv 1(2) \\ \mathbb{Z}/_{2(n-1)} & \text{if } n \equiv 0(2), \text{ and} \end{cases}$$

$$H_1(\Gamma^n;\mathbb{F}_2) = \mathbb{F}_2 \quad \text{if} \quad n \equiv 0(2).$$

Similarly, $H_1(B_n/\mathbb{Z}_n;\mathbb{Z})$ is isomorphic to $\mathbb{Z}/_{n(n-1)}$ and so if $n \equiv 2$ or $3 \bmod 4$, then $H_1(B_n/Z_n;\mathbb{Z}) = \mathbb{Z}/2 \oplus$ odd torsion. By inspection

$$H_1(B_n/Z_n;\mathbb{Z}) \longrightarrow H_1(\Gamma^n;\mathbb{Z})$$

gives an isomorphism on the 2-primary component for $n \equiv 2(4)$ (which is $\mathbb{Z}/2$).

To finish 2.5, we observe that the above paragraph gives that $H^*(\Gamma^n;\mathbb{F}_2) \longrightarrow H^*(B_n/Z_n;\mathbb{F}_2)$ induces an epimorphism on the module of indecomposables: The module of indecomposables of $H^*(B_n/Z_n;\mathbb{F}_2)$ is given by $QH^*(B_n;\mathbb{F}_2) \oplus (\mathbb{F}_2$ concentrated in degree 2). The class in degree 2 is in the image of Sq^1 by the above paragraph and [7, p. 347].

5. PROOFS OF 2.6 AND 2.7

To describe $D:B_{2g} \to \Gamma^0_{g,1}$ and the resulting operations, it is convenient to use results of B. Wajnryb [13]. He gives the following presentation for $\Gamma^0_{g,1}$: There are generators $a_1, b_1, \ldots, a_n, b_n, d$ with relations

(i) $a_ib_ia_i = b_ia_ib_i$ and $a_{i+1}b_ia_{i+1} = b_ia_{i+1}b_i$

for all i,

(ii) $b_2db_2 = db_2d$,

(iii) every pair generators not listed in (i) and (ii) commute

(iv) $(a_1b_1a_2)^4 = d(b_2a_2b_1a_1a_1b_1a_2b_2)^{-1}db_2a_2b_1a_1a_1b_1a_2b_2$, and

(v) $dt_2dt_2^{-1}t_1t_2d(a_1a_2a_3t_1t_2)^{-1} = (ub_1a_2b_2a_3b_3)^{-1}vub_1a_2b_2a_3b_3$

where $t_1 = b_1a_1a_2b_1$, $t_2 = b_2a_2a_3b_2$, $u = a_3b_3t_2d(a_3b_3t_2)^{-1}$, and $v = a_1b_1a_2b_2d(a_1b_1a_2b_2)^{-1}$.

Elements a_i, b_i, and d can be interpreted as Dehn twists with respect to the curves α_i, β_i, and δ below in

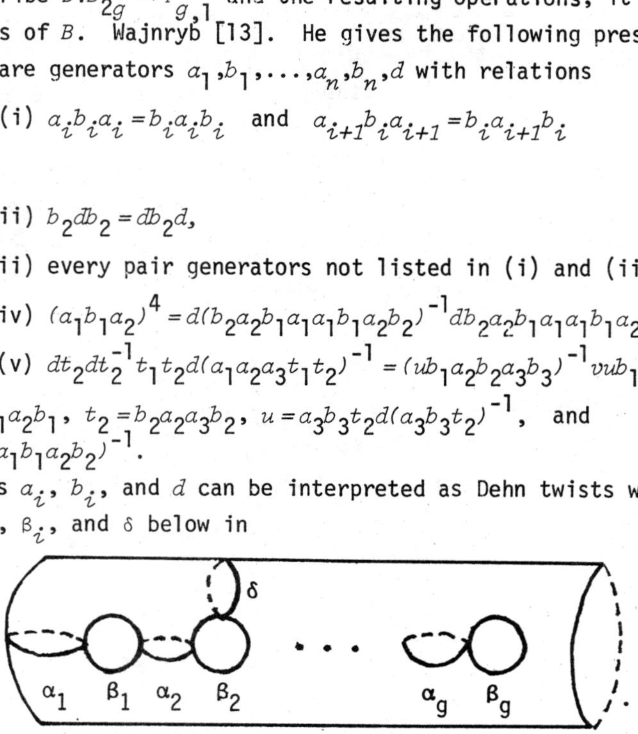

The map D is defined by setting

$$D(\sigma_i) = \begin{cases} b_{\frac{i+1}{2}} & \text{if } i \equiv 1(2) \text{ and} \\ a_{\frac{i}{2}+1} & \text{if } i \equiv 0(2). \end{cases}$$

Evidently D is a homomorphism.

Recall that the wreath product $B_p \wr G$ is $B_p \times G^p$ as a set. Let $x = (b;g_1, \ldots, g_p)$ and $y = (c;h_1, \ldots, h_p)$ be in $B_p \wr G$ with $b, c \in B_p$ and $g_i, h_j \in G$. Then the product xy is defined to be

$$(bc\,;g_{\sigma(1)}{}^{h_1}, g_{\sigma(2)}{}^{h_2}, \dots, g_{\sigma(p)}{}^{h_p})$$

where σ in Σ_p is given by $\sigma=\pi(c)$ and $\pi\colon B_p \to \Sigma_p$ is the natural quotient map.

Maps $\theta'\colon B_p \int B_{2g} \to B_{2pg}$ and $\theta\colon B_p \int \Gamma^0_{g,1} \to \Gamma^0_{gp,1}$ will be given such that the following diagram

$$\begin{array}{ccc} B_p \int B_{2g} & \xrightarrow{\theta'} & B_{2pg} \\ (*)\quad B_p\int\downarrow D & & \downarrow D \\ B_p \int \Gamma^0_{g,1} & \xrightarrow{\theta} & \Gamma^0_{pg,1} \end{array}$$

commutes. To define θ and θ', it suffices to give $\Theta(b;x\,1, \dots, 1)$ where x is in $\Gamma^0_{g,1}$ or B_{2p}. That the correct relations are satisfied is left to the reader.

Define

$$\theta(1\,;x\,,1\,,\dots,1) = x \quad \text{and}$$

$$\theta(\sigma_i;1\,,\dots,1) = x_{1,i}y_{1,i}\cdots x_{g,i}y_{g,i}$$

where

$$y_{j,i} = b_{ig-j+1}a_{ig-j+2}b_{ig-j+2}a_{ig-j+3}\cdots b_{(i+1)g-j}\ \ a_{(i+1)g-j+1} \quad \text{and}$$

$$x_{j,i} = a_{ig-j+2}b_{ig-j+2}a_{ig-j+3}b_{ig-j+3}\cdots\ a_{(i+1)g-j+1}b_{(i+1)g-j+1}$$

Define

$$\theta'(1\,;x\,,1\,,\dots,1) = x \quad \text{and}$$

$$\theta'(\sigma_i\,;1\,,\dots,1) = v_{1,i}v_{2,i}\cdots v_{2g,i}$$

where

$$v_{j,i} = \sigma_{2gi-j+1}\sigma_{2gi-j+2}\cdots\sigma_{(2g)(i+1)-j}.$$

A direct check gives that θ and θ' induce homomorphisms and that diagram (*) commutes. It is worthwhile tracing these homomorphisms through Dehn twists.

Next observe that $H_1\Gamma^0_{g,1} = 0$ if $g \geq 3$ by a check of the relations for $\Gamma^0_{g,1}$. Thus $D\colon H_1(B_{2g};\mathbb{F}_p) \to H_1(\Gamma^0_{g,1};\mathbb{F}_p)$ is zero. But by [7; pp. 347-349], $H_*(B_{2gk};\mathbb{F}_p)$ for $k >> *$ and $*$ increasing with k is generated by operations on $H_1(B_{2g};\mathbb{F}_p)$ together with products. Thus D_* is trivial and Corollary 2.7 follows.

BIBLIOGRAPHY

1. V.I. Arnol'd, "Cohomology of the group of dyed braids," Math. Zametki, 5 (1969), No. 2, 227-231.

2. E. Artin, "Theory of braids," Ann. of Math., 48 (1947), 101-126.

3. J. Birman, "Braids, Links, and Mapping Class Groups," Ann. of Math. Studies, V. 82, Princeton Univ. Press, 1975.

4. J. Birman, and H. Hilden, "Isotopies of homeomorphisms of Riemann surfaces," Ann. of Math., 97 (1973), 424-439.

5. W.L. Chow, "On the algebraic braid group," Ann. of Math., 49 (1948), 654-658.

6. F.R. Cohen, Artin's braid group and classical homotopy theory, Contemp. Math, 44 (1985), 207-220.

7. F.R. Cohen, T.J. Lada, and J.P. May, The Homology of Iterated Loop Spaces, Springer-Verlag, L.N.M., V. 533, New York, 1976.

8. Z. Fiedorowicz and S. Priddy, Homology of Classical Groups Over Finite Fields and Their Associated Infinite Loop Spaces, Springer-Verlag, L.N.M., V. 674, New York, 1978.

9. R. Lee and S. Weintraub, Cohomology of $Sp_4(\mathbb{Z})$ and related groups and spaces, to appear.

10. W. Magnus, "Über Automorphismen von Fundamentalgruppen Berandeter Flächen," Math. Annalen 109 (1934), 617-646.

11. S. Mitchell and S. Priddy, "Stable splittings derived from the Steinberg module," Topology 22 (1983), 285-298.

12. J. Powell, "Two theorems on the mapping class group of surfaces," Proc. Amer. Math. Soc. 68 (1978), 347-350.

13. B. Wajnryb, "A simple presentation for the mapping class group of an orientable surface", Israel J. Math., 45 (1983), 157-174.

THE UNIVERSITY OF KENTUCKY
LEXINGTON, KENTUCKY 40506

Contemporary Mathematics
Volume **58**, Part II, 1987

IMMERSIONS OF REAL PROJECTIVE SPACES

Donald M. Davis[1]

ABSTRACT. This expository paper discusses a strong nonimmersion theorem for real projective spaces. A sketch of the proof, which is quite elementary, is given. A rather detailed comparison of this result with other known results is presented.

1. INTRODUCTION. The question of finding the smallest Euclidean space in which each real projective space P^n can be immersed has attracted the attention of many algebraic topologists since the pioneering work of H. Whitney around 1940. ([23]) Its appeal stems from its being a natural question in differential topology which lends itself to a variety of methods of algebraic topology with amusing numerical results, generally involving the function $\alpha(n)$, the number of 1's in the binary expansion of n.

Best possible results are known for $n \leq 23$ and for $2^i \leq n \leq 2^i + 7$; the first unknown question is whether P^{24} can be immersed in R^{38}. However, prior to the work discussed here, there were arbitrarily long gaps in which no nonimmersion results were known; i.e., there were integers n and g with g arbitrarily large such that for all m between n and $n+g$ the best known nonimmersion of P^m was in the Euclidean space in which P^n was known not to immerse.

The author recently proved the following result. ([11])

THEOREM 1.1. *$P^{2(m+\alpha(m)-1)}$ cannot be immersed in $R^{4m-2\alpha(m)}$.*

The best nonimmersion for each P^{2n} implied by 1.1 is easily seen to be given by the following result.

COROLLARY 1.2. *If d is the smallest nonnegative integer such that $\alpha(n-d) \leq d + 1$, then P^{2n} cannot be immersed in $R^{4n-4d-2\alpha(n-d)}$.*

This result is very strong in several senses which will be explored more fully in Section 5. It has no gaps containing more than one even value of n;

1980 Mathematics Subjection Classification. 57R42, 55N20
[1]Supported by National Science Foundation.

i.e., the nonimmersion for P^{2n+4} implied by 1.2 is always in a larger Euclidean space than that for P^{2n}. It improves upon previous nonimmersion results by arbitrarily large amounts for certain P^n. And it is in a certain sense within two dimensions of all known nonimmersions. In Section 5, which can be read independently of the other sections, we also present a graph comparing various immersion and nonimmersion results.

The other nice thing about Theorem 1.1 is that is proof is surprisingly simple when compared with most other approaches to the immersion question. It can serve as a prototype for arguments in algebraic topology, for it just involves a standard argument that associated to an immersion would be a certain ring homomorphism, and then some linear algebra to show such a homomorphism cannot exist. This homomorphism is the cohomology homomorphism induced by an axial map; the only novelty is that we use an "extraordinary" cohomology theory whose definition is quite complicated, but whose properties are quite ordinary.

In Section 6, we offer for comparison known results on immersions of complex projective spaces, lens spaces, and all manifolds.

2. SOME TOPOLOGICAL BACKGROUND. *Real projective space* P^n can be defined as the set of lines through the origin in R^{n+1}, or as the space obtained from the n-sphere S^n by identifying antipodal points $(x \sim -x)$. The former description shows its connection with projective geometry, but the latter is more useful for our purposes. Both S^n and P^n are (differentiable) *n-manifolds*, topological spaces which are locally homeomorphic to R^n in such a way that those homeomorphisms which overlap do so in a differentiable way. An *immersion* of an n-manifold M in R^k is a function $M \to R^k$ which at each point of M sends the n-dimensional vector space of tangent vectors to M at x injectively to R^k. An immersion which is globally injective is an *embedding*. The familiar picture of the Klein bottle K as a cylinder with the ends identified in such a way that the cylinder must pass through itself in order to make the identification is of an immersion in R^3; K cannot be embedded in R^3.

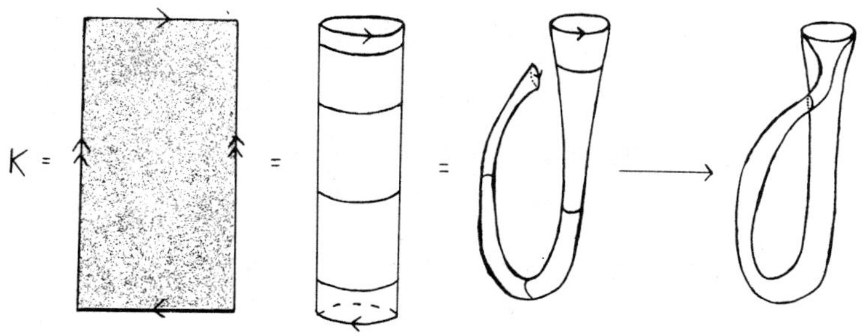

We paraphrase an argument of [19] that if P^n immerses in R^{n+k} then there is a nonsingular skew map $f: S^n \times S^n \to R^{n+k+1} - 0$, $(f(-x,y) = -f(x,y) = f(x,-y))$, which necessarily passes to an axial map $P^n \times P^n \to P^{n+k}$, a map whose restriction to each P^n is nontrivial. The tangent vectors to S^n at y are those v in R^{n+1} such that the inner product $\langle v,y\rangle = 0$. If we identify (v,y) with $(-v,-y)$, the space of tangent vectors of P^n is obtained. The immersion induces a map

$$j: \{(v,y) \in R^{n+1} \times S^n: \langle v,y\rangle = 0\} \to R^{n+k}$$

which is a monomorphism for fixed y, and satisfies $j(-v,-y) = j(v,y)$.

Define $f: R^{n+1} \times S^n \to R^{n+k+1}$ by

$$f(x,y) = (\langle x,y\rangle, j(x-\langle x,y\rangle y,y)).$$

The restriction to $S^n \times S^n$ is a nonsingular skew map.

3. SKETCH OF PROOF OF 1.1. We will sketch the proof of

THEOREM 3.1. *There is no axial map*

$$P^{2(m+\alpha(m)-1)} \times P^{2(m+\alpha(m)-2)} \to P^{4m-2\alpha(m)-1}.$$

By Section 2, this implies a nonimmersion result 1 dimension weaker than 1.1. To get the extra dimension requires a modification of the method of Section 2.

Let $\nu(n)$ denote the exponent of 2 in the prime factorization of n. There is a generalized cohomology theory $B^*(\ )$, which will be described in more detail in Section 4, which satisfies

PROPOSITION 3.2. *i) If X is a topological space, then $B^*(X)$ is a graded ring, i.e. for all integers i, $B^i(X)$ is an abelian group and there is an associative pairing $B^i(X) \otimes B^j(X) \to B^{i+j}(X)$.*

ii) If $f: X \to Y$ is a continuous map of topological spaces, there is a homomorphism of graded rings $f^*: B^*(Y) \to B^*(X)$.

iii) There is an element $X \in B^2(P^c)$ such that $X^i = 0$ if $2i > c$.

iv) For any a and b, there are elements X_1 and X_2 in $B^2(P^a \times P^b)$ corresponding to the factors.

v) For any n and k, let $s = \nu\binom{n}{k}$. Then

$$(X_1-X_2)^{n-s} \neq 0 \text{ in } B^{2(n-s)}(P^{2(k+s)} \times P^{2(n-k+s)}).$$

vi) If $f: P^a \times P^b \to P^c$ is an axial map, then $f^*(X) = u\cdot(X_1-X_2)$, where u is a unit in $B^*(P^a \times P^b)$.

Most of our work goes into proving (v). A result similar to (vi) was proved by Astey in [1], which was a precursor of this work.

We now deduce 3.1 from 3.2. It is easy to see that $\nu\binom{2m-1}{m} = \alpha(m)-1$. If the axial map of 3.1 exists,

$$0 = f^*(X^{2m-\alpha(m)}) = (u\cdot(X_1-X_2))^{2m-\alpha(m)}$$

by (iii) and (vi). Since u is a unit, this contradicts (v) with $n = 2m-1$ and $k = m$.

4. B^*-COHOMOLOGY GROUPS. The argument of Section 3 for nonexistence of axial maps is quite similar to one used by Hopf in [13] using ordinary cohomology theory $H^*(\ )$. Our generalized cohomology theory $B^*(\ )$ is formally similar except that the groups tend to be larger and involve negative (as well as positive) indices.

Complex cobordism cohomology theory $MU^*(\ )$ was one of the first generalized cohomology theories when it was introduced in [3]. It had arisen from the study of classification of complex manifolds under the relation of being boundaries of higher dimensional manifolds, but could be formally viewed similarly to 3.2. For many purposes, it gives more terms than one can conveniently consider, which was simplified somewhat by the Brown-Peterson cohomology theory $BP^*(\ )$ ([6],[7]), which localizes $MU^*(\ )$ at a prime and removes extraneous terms. Calculations with $BP^*(\ )$ involve terms v_i for all positive integers i, and for a product of 2 real projective spaces only the interplay of v_1 and v_2 are relevant. There is a cohomology theory $BP\langle 2\rangle^*(\ )$, which we use here abbreviated to $B^*(\ )$, constructed using manifolds with singularities in [4], which contains exactly this information.

Theorem 3.2(v) follows by rather straightforward linear algebra from

THEOREM 4.1. *In even degrees greater than* $2k+6$ *and* $2m+6$, $B^*(P^{2k}\times P^{2m})$ *is the quotient of the polynomial ring* $\mathbf{Z}_{(2)}[X_1,X_2]$ *by the ideal generated by* $2^i X_2^{m+1-i}$ *and* $2^i X_1^{k+1-i}/(X_1-X_2)^i$ *for* $i \geq 0$.

Here $\mathbf{Z}_{(2)}$ denotes the integers localized at 2, consisting of rational numbers with odd denominators. In the rest of this section we discuss some of the ideas in the proof of 4.1.

For any space X, the cohomology ring $B^*(X)$ is a module over $B^* = \mathbf{Z}_{(2)}[v_1,v_2]$, where v_1 and v_2 have degrees -2 and -6, respectively. A formal description of the groups described more explicitly in 4.1 was well-known (see e.g. [1]).

PROPOSITION 4.2. *In even degrees greater than* $2k+6$ *and* $2m+6$

$$B^*(P^{2k}\times P^{2m}) \approx B^*[X_1,X_2]/(X_1^{k+1},X_2^{m+1},[2](X_1),[2](X_2)).$$

What makes this cumbersome is the presence of the 2-series $[2](X)$, which is a power series in X with coefficients in B^*. It is computable, but its calculation is very tedious, even for a computer. It begins

$$2X - v_1X^2 + 2v_1^2X^3 - (7v_2+8v_1^3)X^4 + \ldots$$

It turns out that all we need to know about it is that mod 2 it is a series in X^2, which follows from [24; 3.17].

The series $[2](X)$ can be multiplied by a unit to yield a "reduced" series which allows one to eliminate v_1 from 4.2 in 2 different ways, corresponding to X_1 and X_2. Comparing these yields the following description.

PROPOSITION 4.3. *In even degrees greater than* $2k + 6$ *and* $2m + 6$

$$B^*(P^{2k}\times P^{2m}) \approx \mathbf{Z}_{(2)}[X_1,X_2,v_2]/(X_1^{k+1},X_2^{m+1},\Sigma\alpha_j v_2^j(X_1^{3j-1}-X_2^{3j-1}))$$

where α_j *are coefficients in the reduced 2-series satisfying* α_j *is even if* j *is even.*

We form a matrix presentation for $B^{2k+2m-2d}(P^{2k}\times P^{2m})$, involving monomials in v_2, X_1, and X_2. This matrix is independent of k and m (in the range of 4.3). Row reduction of this matrix allows one to express monomials divisible by v_2 in terms of those not divisible by v_2, and yields relations among the monomials in X_1 and X_2 which can be interpreted as 4.1.

5. SOME NUMERICAL EXAMPLES. In this section we give some numerical examples of 1.1 and 1.2, and discuss their strength by comparing with previously known results. Denote the dimension of the largest Euclidean space in which P^n is proved not to immerse by

K(n) for all results except Corollary 1.2.

D(n) for Corollary 1.2.

Let I(n) denote the dimension of the smallest Euclidean space in which P^n is known to immerse.

A short table is given below to illustrate the way that only certain values of m in 1.1 are useful, and how these are related to d of 1.2. It also compares our results with K(n) in a range (near a 2-power) where previous methods have been most successful.

n	m	d	D(n)	K(n)	I(n)
124	58	4	224	231	238
125			224	231	238
126	60	3	232	232	238
127			232	238	240
128	64	0	254	254	255
129			254	254	255
130	64	1	254	255	256
131			254	255	256
132	65	1	256	257	258
133			256	261	262
134	66	1	260	261	262
135			260	261	262
136	66	2	260	261	263
137			260	269	270
138	68	1	268	269	270
139			268	269	270
140	68	2	268	270	271
141			268	270	273
142	69	2	270	270	275

Our result is within 2 dimensions of previous results in the sense that we allow the 2 dimensions to come from the Euclidean space or the projective space, or to be split between them.

PROPOSITION 5.1. *For all* n, *either* $D(n) \geq K(n) - 2$
or $D(n+1) \geq K(n) - 1$
or $D(n+2) \geq K(n) + 1$.

This is verified by perusal of the literature (see [10], [14], and [18]). The "+1" in the third case is just for good measure. The first type of inequality is necessary for $n = 140$, the second for 133, and the third for 124.

It seems remarkable that such a simple statement as 1.1 with such a simple proof could come so close to all results proved by a wide variety of more complicated methods. It seems reasonable to venture the conjecture that in this sense our result may be close to best possible, i.e. that 5.1 is true with $K(n)$ replaced by the dimension of the largest Euclidean space in which P^n does not immerse, but a proof of the necessary immersion results does not appear to be near at hand.

As we remarked in the introduction, previous nonimmersion results had arbitrarily long gaps in which no new results were known. For example,

$$K(2^{2^k+k+2}-2^{k+2}) = \ldots = K(2^{2^k+k+2}-2^{k+1}-3) = 2^{2^k+k+3}-7\cdot 2^{k+1} \quad ([2])$$

Our results do not suffer from this defect:

PROPOSITION 5.2. $D(N+4) > D(N)$ *for all* N.

Proof. If 1.1 implies that P^{2n} does not immerse in $R^{4m-2\alpha(m)}$, then it also implies $P^{2(n+2)}$ does not immerse in $R^{4(m+1)-2\alpha(m+1)} > 4m-2\alpha(m)$. This is true because $\alpha(m+1) \leq \alpha(m) + 1$. ∎

In particular, we have arbitrarily large improvement over K( ). For example, in the middle of the above gaps, we have

$$D(2^{2^k+k+2}-3\cdot 2^k) = 2^{2^k+k+3}-6\cdot 2^{k+1}-2,$$

improving upon K( ) by $2^{k+1}-2$ dimensions. (Here d in 1.2 will equal 2^k.)

The gaps in known nonimmersion results were much longer prior to Astey's thesis in 1978. In it, he initiated the application of BP to the immersion question ([1],[2]). Our approach is topologically equivalent to his; we were just lucky to find that a little extra work with the algebra led to a major improvement in the results.

The graphs on the next page illustrate these relations for immersions and nonimmersions of P^n for $7700 \leq n \leq 8300$. (n.b. $2^{13} = 8192$). The five functions graphed are the functions K(n), D(n), and I(n) defined above, Cohen's function $C(n) = 2n - \alpha(n)$ (see 6.1), and the pre-Astey nonimmersion results P(n). Of course we have $I(n) > K(n) \geq P(n)$ and $I(n) > D(n)$. Theorem 6.1 or the comments after it imply $C(n) \geq I(n)$. Not so obvious are the following

i) $D(n) \geq K(n)$ for $7690 \leq n < 8190$

ii) $K(n) \geq D(n)$ for $8191 \leq n \leq 9191$.

The main reason for (ii) is that Theorem 1.1 was proved for $\alpha(m) \leq 7$ in [5], where 1.1 was conjectured.

Two different graphs are presented, using different scales, with one graph a subset of the other. In the lower graph, the solid curve is D(n), and the wavy curve above it is I(n). The top dotted curve is C(n). The graph shows that the difference C(n)-I(n) is usually substantial. K(n) is often equal to D(n); for such n, the graph of K is subsumed in the graph of D. For $n < 8191$, the dotted ledges below the D-curve are values of K(n) when $K(n) < D(n)$. The longest of these ledges, occurring at height 16272 is the case $k = 3$ of the long gaps mentioned earlier in this section. For $n \geq 8191$ there are a few stray dots and ×'s slightly above the D-curve, which are the cases $K(n) > D(n)$. If they are ×'s, they are also values of P(n), while if they are dots $K(n) > P(n)$. The ledges of little ×'s below the D-curve for $n \geq 8191$ are values of P(n) when $P(n) < D(n)$.

Now for P(n) when $n < 8191$: This is the reason for the two graphs.

$$P(7687) = \ldots = P(8190) = 15354. \quad ([9])$$

This is so far below K(n) and D(n) that it can only be shown on a scale in which differences between K(n), D(n), etc., are not so clear. In this (upper) graph, the long ×'d line is P(n). The two straight lines drawn lightly above the domain $7686 \leq n < 8000$ are crude approximations to I(n) and D(n). The precise values of the functions at the left end of this graph are

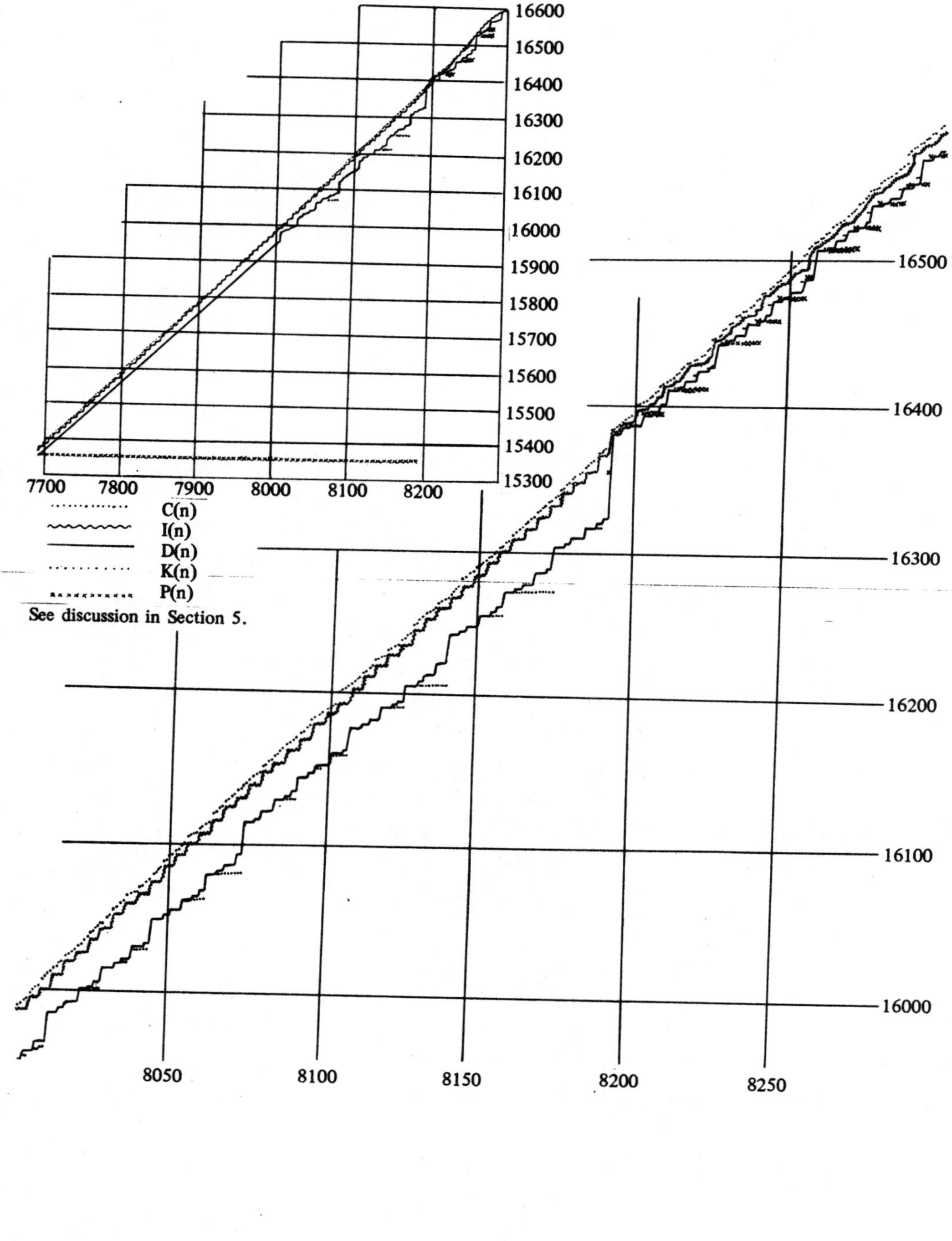

See discussion in Section 5.

n	P(n)	K(n)	D(n)	I(n)	C(n)
7686	14842	15352	15352	15360	15366
7687	15354	15354	15352	15360	15367
7688	15354	15354	15352	15367	15371
7689	15354	15354	15352	15367	15372
7690	15354	15354	15354	15367	15374

6. SOME RELATED RESULTS. In this section we mention results on three immersion questions: for manifolds, complex projective spaces, and lens spaces.

In a highly celebrated result, Ralph Cohen proved

THEOREM 6.1 ([7]). *Every n-manifold can be immersed in* $R^{2n-\alpha(n)}$.

The techniques used to prove 6.1 are much deeper than those of this paper, both because it deals with *all* manifolds and because immersions are more difficult to prove than nonimmersions.

6.1 is best possible, because if $n = \Sigma 2^{e_i}$ corresponds to the binary expansion of n (i.e. e_i are distinct), then $\prod P^{2^{e_i}}$ is an n-manifold which does not immerse in $R^{2n-\alpha(n)-1}$. Cohen's result gave no new information for real projective spaces because Milgram had proved much earlier ([16]) that P^n immerses in $R^{2n-\alpha(n)}$.

The immersion question for complex projective spaces CP^n is much better understood than that for real projective spaces. The following result, which was proved by various applications of K-theory, is probably within 1 of best possible.

THEOREM 6.2 ([15],[20],[12]). CP^n *cannot be immersed in* $R^{4n-2\alpha(n)+\varepsilon}$, *where*

$$\varepsilon = \begin{cases} 0 & \text{if } n \text{ is even and } \alpha(n) \equiv 1 \ (4) \\ 1 & \text{if } n \text{ is even and } \alpha(n) \equiv 2 \text{ or } 3 \ (4) \\ -1 & \text{otherwise} \end{cases}$$

In [12] immersions of CP^n for $\alpha(n) < 8$ were obtained in Euclidean spaces 1 or 2 dimensions greater than those of 6.2.

The borderline cases were given careful attention in [21], with some new nonimmersions deduced, but in a much less tractable form. The method again is K-theory.

THEOREM 6.3 ([21]). *If* CP^n *immerses in* $R^{4n-2\alpha(n)}$, *then*

$$\nu(c_{\alpha(n)-1}) = \nu(c_{\alpha(n)}) < \nu(c_i) \quad \text{for all } i < \alpha(n)-1,$$

$$\text{where } \Sigma c_i t^i = ((\log(1+t))/t)^{2n+1-\alpha(n)}.$$

Recently, Crabb ([8]) appears to have shown that the nonimmersions of 6.2 and 6.3 are best possible for $\alpha(n) < 7$ by establishing corresponding immersions, using an elaborate combination of K-theory and obstruction theory.

A natural generalization of P^{2n+1} is the *lens space* $L^{2n+1}(p)$ associated to the prime p which is the quotient of $S^{2n+1} \subseteq C^{n+1}$ by the relation $(z_0,\ldots,z_n) \sim (cz_0,\ldots,cz_n)$ if $c^p = 1$ (i.e. $c = e^{k2\pi i/p}$ for some integer k).

It is a (2n+1)-manifold.

Good results have been obtained for immersions of lens spaces by using obstruction theory.

THEOREM 6.4 ([22]). *Let p be odd.*

i) $L^{2n+1}(p)$ *can be immersed in* $R^{2n+2[n/2]+2}$;

ii) if $p > [n/2]+2$, *then* $L^{2n+1}(p)$ *can be immersed in* $R^{2n+2[n/2]+1}$ *iff* $(-1)^{[n/2]}\binom{n+[n/2]}{n}$ *is a quadratic residue mod* p;

iii) if $\binom{n+[n/2]}{n} \not\equiv 0$ *(p), then* $L^{2n+1}(p)$ *cannot be immersed in* $R^{2n+2[n/2]}$.

This gives a complete answer if $p > n+[n/2]$.

BIBLIOGRAPHY

1. L. Astey, "Geometric dimension of bundles over real projective spaces," Quar. J. Math Oxford 31 (1980) 139-155.

2. ________ and D.M. Davis, "Nonimmersions of real projective spaces implied by BP," Bol. Sox Mat Mex 25 (1980) 15-22.

3. M.F. Atiyah, "Bordism and Cobordism," Proc Camb Phil Soc 57 (1961) 280-288.

4. N.A. Baas, "On bordism theory of manifolds with singularities," Math Scand 33 (1973) 279-302.

5. M. Bendersky and D.M. Davis, "Unstable BP-homology and desuspensions," to appear in Amer Jour Math.

6. E.H. Brown and F.P. Peterson, "A spectrum whose $\mathbb{Z}_p$-cohomology is the algebra of reduced p^{th}-powers," Topology 5 (1966) 149-157.

7. R. Cohen, "Immersions of manifolds," to appear in Annals of Math.

8. M.C. Crabb, "On the $KO_{\mathbb{Z}/2}$-Euler class", preprint.

9. D.M. Davis, "Connective coverings of BO and immersions of projective spaces," Pac Jour Math 76 (1978) 33-42.

10. ________, "Some new immersions and nonimmersions of real projective spaces," Proc. Northwestern Homotopy Theory Conf, Contemporary Math, Amer Math Soc 19 (1982) 51-64.

11. ________, "A strong nonimmersion theorem for real projective spaces," Annals of Math. 120 (1984) 517-528.

12. ________ and M. Mahowald, "Immersions of complex projective spaces and the generalized vector field problem," Proc London Math Soc 35 (1977) 333-348.

13. H. Hopf, "Ein Topologischer Beitrag zur reellen Algebra," Comm Math Helv 13 (1941) 219-239.

14. I.M. James, "Euclidean models of projective spaces," Bull London Math Soc 3 (1971) 257-276.

15. K.H. Mayer, "Elliptische Differentialoperatoren und Ganzzahligkeitssatze fur charakteristische Zahlen," Topology 4 (1965) 295-313.

16. R.J. Milgram, "Immersing projective spaces," Annals of Math 85 (1967) 473-482.

17. D. Quillen, "On the formal group laws of unoriented and complex cobordism theory," Bull Amer Math Soc 75 (1969) 1293-1298.

18. D. Randall, "On equivariant maps of Stiefel manifolds," to appear.

19. B. Sanderson, "A nonimmersion theorem for real projective spaces," Topology 2 (1963) 209-211.

20. ____________ and R. Schwarzenberger, "Nonimmersion theorems for differentiable manifolds," Proc Camb Phil Soc 59 (1963) 312-322.

21. F. Sigrist and U. Suter, "On immersions of CP^n in $R^{4n-2\alpha(n)}$," Proc Vancouver Conf, Springer Verlag Lecture Notes in Math, 673 (1978).

22. D. Sjerve, "Geometric dimension of vector bundles over lens spaces," Trans Amer Math Soc 134 (1968) 545-558.

23. H. Whitney, "The singularities of a smooth n-manifold in (2n-1)-space," Annals of Math 45 (1944) 247-293.

24. W.S. Wilson, *A BP-introduction and sampler,* CBMS Regional Conf Series 48 (1982).

DEPARTMENT OF MATHEMATICS
LEHIGH UNIVERSITY
BETHLEHEM, PA 18015

Contemporary Mathematics
Volume **58**, Part II, 1987

FRACTAL STRUCTURES IN $H_*(BO)$ AND THEIR APPLICATION TO COBORDISM

Vincent Giambalvo, David J. Pengelley and Douglas C. Ravenel *

The mod 2 homology of BO, the classifying space for real vector bundles, has been well understood for a long time. We have had explicit descriptions of its Hopf algebra structure and of the action of the Steenrod algebra on it for 20 years or more. Yet it is not without its surprises. We will describe a hitherto unnoticed structure in it which is very useful for certain problems in cobordism theory.

This structure is a grading, independent of the usual topological grading, which sheds new light on the action of the Steenrod algebra. It is described in Section 1. Our proof is very computational and gives no insight into the true origin of this phenomenon, which remains a mystery to us. The details, which are completely supressed here, will appear in [GPR]. In Section 2 we will describe the application of our result to cobordism theory. In Section 3 we will explain our use of the term "fractal" in the title.

1. A decomposition of $H_*(BO)$.

This work began with the following observation by the third author in connection with the work of the first two [GP] on Spin cobordism. To study the latter one computes $\pi_*(M\mathrm{Spin})$ with the Adams spectral sequence, whose E_2-term is

$$\mathrm{Ext}_A(Z/2,\ H_*(M\mathrm{Spin}))$$

where A is the mod 2 Steenrod algebra and the homology has coefficients in $Z/2$. The Ext here and elsewhere in this paper is understood to be in the category of comodules over the dual of A.

From the work of Anderson, Brown and Peterson [ABP] one knows that the A-module generated by the Thom class in $H^*(M\mathrm{Spin})$ is A/ASq^1+ASq^2. It follows by a change-of-rings argument that the Ext group above is isomorphic to

$$\mathrm{Ext}_{A_1}(Z/2,\ N)$$

where A_1 is the subalgebra of A generated by Sq^1 and Sq^2, $N = H_*(M\mathrm{Spin})/I$ and I is a certain ideal. Thus one needs to understand the structure of this N as an algebra over A_1. This problem is analyzed in [GP].

* *The third author was partially supported by the NSF.*

The result is that there is a tensor product decomposition

$$N = P \otimes \otimes_{i \geq 1} N(i)$$

as A_1-algebras. P is a polynomial algebra

$$P = Z/2[y_8, y_{16}, y_{32}, \cdots]$$

with trivial A_1-action where dim $y_m = m$. The factors $N(i)$ are all isomorphic *up to regrading.*

By this we do *not* mean that they are suspensions of each other. Each $N(i)$ is a polynomial algebra with at most one generator in each dimension. In $N(1)$ these generators occur in dimensions 10 through 14, 23 through 30, 47 through 62, etc. In $N(2)$ they occur in dimensions 18 through 22, 39 through 46, 79 through 94, etc. In each case generators in the first block of dimensions are connected to each other by the A_1-action, i.e. by the action of Sq^1 and Sq^2. Generators in the second block are connected to each other and to two-fold products of elements in the first block, and so on.

These results about N were all proved in [GP].

The observation that led to the present work is that each of these factors $N(i)$ is also isomorphic as an A_1-algebra (up to dimension regrading) to $H_*(BSO)$.

Thus the computation needed to find $\pi_*(M\text{Spin})$ is essentially reduced to the analysis of $H_*(BSO)$ as an A_1-algebra. This leads one to ask if $\pi_*(MO\langle 8\rangle)$ is related in a similar way to $H_*(B\text{Spin})$ as an A_2-algebra, where A_2 is the subalgebra of A generated by Sq^1, Sq^2 and Sq^4. Theorem 2 below is an affirmative answer to this question.

The regrading discussed above comes about in the following way. We will define a new grading on $H_*(BSO)$, independent of the usual topological grading, having the property that if x is a homogeneous element then $Sq^1 x$ and $Sq^2 x$ are also homogeneous and have the same grade. Thus our new grading gives a direct sum decomposition of $H_*(BSO)$ as an A_1-module and turns it into a bigraded algebra. We can regrade it by taking various linear combinations of these two gradings. The $N(i)$ are each obtained by taking the sum of the topological grading and a suitable power of two times our new grading.

Recall that $H_*(BO)$ is a polynomial algebra with one generator in every dimension. We need to consider a sequence of Hopf-subalgebras

$$H_*(BO) = B_0 \leftarrow B_1 \leftarrow B_2 \leftarrow \cdots$$

where $B_1 = H_*(BSO)$, $B_2 = H_*(B\text{Spin})$, $B_3 = H_*(BO\langle 8\rangle)$, $B_4 = imH_*(BO\langle 9\rangle)$ and B_n is the image of the homology of the nth distinct connected cover $BO\langle\phi(n)\rangle$ (not to be confused with the n-connected cover) of BO.

Here $BO\langle\phi(n)\rangle$ denotes the $(\phi(n)-1)$-connected cover of BO, where $\phi(n)$ denotes the dimension of its $(n+1)$th nontrivial homotopy group. Thus we have $\phi(1)=2$, $\phi(2)=4$, $\phi(3)=8$, $\phi(4)=9$, etc. B_n and $H_*\big(BO\langle\phi(n)\rangle\big)$ coincide only for $n\leq 3$. The function $\phi(n)$ is roughly $2n$ while the connectivity of B_n is 2^n.

These algebras were studied by Baker [B] and Kochman [K]. Their structure is given by

Lemma. *$H_*(BO)$ has polynomial generators x_i such that*

$$B_n = Z/2[x_i^{e(n,i)} : i > 0]$$

where $e(n,i)=2^m$, $m=\max\{0,n+1-\alpha(i)\}$ and $\alpha(i)$ is the number of ones in the binary expansion of i. ■

For example

$$\begin{aligned} B_2 &= H_*(B\mathrm{Spin}) \\ &= Z/2[y_4, y_6, y_7, y_8, y_{10}, \cdots] \end{aligned}$$

where

$$\begin{aligned} y_4 &= x_1^4, \\ y_6 &= x_3^2, \\ y_7 &= x_7, \\ y_8 &= x_2^4 \quad \text{and} \\ y_{10} &= x_5^2. \end{aligned}$$

The x_i are essentially the generators given by Husemoller in [H]. In particular x_i is primitive whenever i is odd.

Now the Steenrod algebra A has subalgebras

$$A_0 \to A_1 \to A_2 \to \cdots$$

where A_n is the subalgebra generated by

$$\{Sq^i : i \leq 2^n\}.$$

Our first result is

Theorem 1. *$H_*(BO)$ has a grading compatible with its algebra structure such that*

(i) when restricted to B_n the grading gives a direct sum decomposition over A_n,

(ii) polynomial generators u_i can be chosen which are homogeneous of grade 2^k for $2^k-1\leq i\leq 2^{k+1}-2$, and

(iii) the description of B_n given in the Lemma above still holds with the x_i replaced by the u_i.

In particular $u_i = x_i$ when $i = 2^k - 1$. ∎

The polynomial generators u_i are given in terms of the x_i by a complicated formula which will not be given here. We will, however, show why the x_i's themselves will not do by making some calculations in low dimensions. The right action of the Steenrod algebra on the x_i was determined by Lance [L]. For example, we have

$$\begin{aligned}(x_2)Sq^1 &= x_1,\\ (x_3)Sq^1 &= x_1^2,\\ (x_3)Sq^2 &= 0,\\ (x_4)Sq^1 &= x_3 + x_2x_1,\\ (x_5)Sq^1 &= x_1^4 \quad\text{and}\\ (x_5)Sq^2 &= x_3.\end{aligned}$$

There is no problem in defining $u_i = x_i$ for $i \le 4$, but we cannot define u_5 to be x_5 for the following reason. The grade of u_5 should be preserved by both Sq^1 and Sq^2, but $(x_5)Sq^1$ has grade 8 while $(x_5)Sq^2$ has grade 4. We can get around this difficulty by defining

$$u_5 = x_5 + x_1^2x_3,$$

which gives

$$\begin{aligned}(u_5)Sq^1 &= 0 \quad\text{and}\\ (u_5)Sq^2 &= u_3,\end{aligned}$$

so u_5 may have grade 4 as desired.

A more geometric question raised by this theorem is the following. The theorem tells us that B_n is the image of $H_*\big(BO\langle\phi(n)\rangle\big)$ in $H_*(BO)$ and it has a nice splitting over A_n. Suppose then that $H_*\big(BO\langle\phi(n)\rangle\big)$ itself has such a splitting, which we know it does for $n \le 3$.

What might this mean geometrically? If there were a spectrum X_n with

$$H^*(X_n) = A \otimes_{A_n} Z/2$$

then we could expect a splitting of $BO\langle\phi(n)\rangle \wedge X_n$.

Unfortunately the spectra X_n are known to exist only for $n = 0$ and $n = 1$. The spectra in question are the integer Eilenberg-MacLane spectrum HZ and the connective real K-theory spectrum *bo*. X_2 is known not to exist by [DM]. For $n \ge 3$ the existence of such an X_n would run afoul of the Hopf invariant one theorem. However, we can use $MO\langle\phi(n+1)\rangle$ as a substitute for X_n

since the A-module generated by the Thom class in $H^*\left(MO\langle\phi(n+1)\rangle\right)$ is the same as $H^*(X_n)$. Thus we have

Question. Is there a splitting of

$$BO\langle\phi(n)\rangle \wedge MO\langle\phi(n+1)\rangle$$

corresponding to that of Theorem 1? ∎

2. Application to cobordism.

Cobordism is an equivalence relation among closed manifolds, in which a manifold is cobordant to the empty set if it is the boundary of another manifold. In cobordism theory one wants to compute the ring of cobordism classes of manifolds with a specified structure, e.g. an orientation, a Spin structure, etc.

These structures always pertain to the stable normal bundle, i.e. the normal bundle for an embedding of the manifold M in a large dimensional Euclidean space. There is always a classifying space B for vector bundles equipped with the structure in question, and M comes equipped with a map to B inducing its stable normal bundle. For more background on this subject see Stong [S].

In the case we are interested in, B is $BO\langle\phi(n)\rangle$ (the $(\phi(n)-1)$-connected cover of BO), so our manifolds come equipped with a trivialization of the stable normal bundle restricted to the $(\phi(n)-1)$-skeleton.

Thom's transversality theorem tells us that the group of cobordism classes of n-dimensional manifolds equipped with the structure at hand is isomorphic to the nth homotopy group of the Thom spectrum associated with B. Thus it converts a geometric problem to a homotopy theoretic one.

In our case this Thom spectrum is denoted by $MO\langle\phi(n)\rangle$. The E_2-term of the relevant Adams spectral sequence is

$$\mathrm{Ext}_A\left(Z/2,\ H_*\left(MO\langle\phi(n)\rangle\right)\right).$$

Recall the subalgebra B_n of $H_*(BO)$ is defined to be the image of the homology of a suitable connected cover of the space BO. Let M_n be the corresponding subalgebra of $H_*(MO)$. Thus we have $M_0 = H_*(MO)$, $M_1 = H_*(MSO)$, $M_2 = H_*(M\mathrm{Spin})$, and $M_3 = H_*(MO\langle 8\rangle)$. There is a change-of-rings isomorphism

$$\mathrm{Ext}_A(Z/2,\ M_n) = \mathrm{Ext}_{A_{n-1}}(Z/2,\ M_n/I_n)$$

where I_n is the following ideal. Let J be the ideal $(x_1, x_3, x_7, x_{15}, \cdots)$ in $H_*(BO)$. Let J_n by the intersection of J and B_n and define I_n to be the corresponding ideal in M_n.

Theorem 2. *Let $N_n = M_n/I_n$. Then as algebras over A_{n-1}*

$$N_n = P_n \otimes \otimes_{i \geq 1} N_n(i).$$

P_n is a polynomial algebra with one generator in each two-power dimension starting with 2^{n+1} with trivial A_{n-1} action. Each $N_n(i)$ is isomorphic up to regrading as an A_{n-1} algebra to B_{n-1}. ∎

For $n = 3$ this gives the structure theorem for M_3 (i.e. for $MO\langle 8\rangle$) mentioned above. It reduces the problem to that of studying $B_2 = H_*(B\mathrm{Spin})$ as an A_2-algebra. We plan to take this up in the near future. Davis [D] has derived the A_1-structure of M_3 with and without killing the ideal I_3. The former structure follows immediately from Theorem 2 and our knowledge [GP] of the A_1-structure of B_2.

3. Fractals.

There are two senses in which the term "fractal" can be applied to our results.

For the first recall that a fractal is typically a subset of R^2 or R^3 with the property that a small portion of it resembles the whole. For example, the Cantor set has arbitrarily small subsets homeomorphic to the entire space even when the embedding in the unit interval is taken into account.

The term is also used to describe subsets (such as the M-set) which appear to have an increasingly rich structure when viewed on a smaller and smaller scale. We are using the term here in this latter sense. *Our grading on $H_*(BO)$ reveals an increasingly rich structure (i.e. decomposition over larger and larger portions of the Steenrod algebra) when attention is restricted to smaller and smaller subalgebras, i.e. to the B_n's.*

Our second use of the term "fractal" involves a reformulation of Theorem 1 in terms of fractal algebra, by which we mean algebraic objects graded over $Z[1/2]$ instead of Z. We first define the fractal Steenrod algebra FA. A admits a (nongraded) endomorphism f that sends Sq^i to $Sq^{i/2}$ if i is even and to 0 if i is odd.

FA is the inverse limit of the system

$$A \xleftarrow{f} A \xleftarrow{f} A \xleftarrow{f} \cdots.$$

Here the grading on the ith copy of A is the usual topological grading divided by 2^{i-1} so that all the maps are homogeneous and the inverse limit is graded over $Z[1/2]$. FA is generated as an algebra by elements which could reasonably be denoted by

$$Sq^{1/2}, Sq^{1/4}, Sq^{1/8}, \cdots$$

in addition to the usual $Sq^1, Sq^2, Sq^4, \cdots$.

FA has a subalgebra FA_0 defined to be the inverse limit of

$$A_0 \xleftarrow{f} A_1 \xleftarrow{f} A_2 \xleftarrow{f} \cdots.$$

It is the subalgebra generated by $Sq^1, Sq^{1/2}, Sq^{1/4}, Sq^{1/8}, \cdots$.

Next we will define a fractal version of $H_*(BO)$. Define a nongraded endomorphism s of $H_*(BO)$ by sending x to x^2. This s sends B_n into B_{n+1}. FB is defined to be the direct limit of

$$B_0 \xrightarrow{s} B_1 \xrightarrow{s} B_2 \xrightarrow{s} \cdots.$$

Again the grading on B_i is the usual topological grading divided by 2^i so that all the maps are homogeneous and the direct limit is graded over $Z[1/2]$. FB is the polynomial algebra generated by the $2^{\alpha(i)-1}$th root of x_i for each positive i. The subset S of the rationals consisting of the dimensions of these generators has some curious features. It is closed, multiplication by two produces a homeomorphism of S with its set of accumulation points, and the nonaccumulation points form an open dense subset, i.e. S is homeomorphic to the complement of an open dense subset.

FB is easily seen to be an algebra over FA and hence over FA_0. Theorem 1 can be reformulated as

Theorem $F1$. *FB has a grading over $Z[1/2]$ compatible with its algebra structure such that*
(i) it gives a direct sum decomposition over FA_0, and
(ii) its restriction to $H_*(BO)$ coincides with the grading of Theorem 1. ∎

References

[ABP] D.W. Anderson, E.H. Brown and F.P. Peterson, The structure of the Spin cobordism ring, *Ann. of Math.* **86**(1967), 271-298.

[B] A. Baker, Husemoller-Witt decompositions and actions of the Steenrod algebra, preprint.

[D] D. Davis, On the cohomology of $MO\langle 8\rangle$, Proc., Symposium on Algebraic Topology in Honor of Jose Adem, *Contemporary Math.* **12**(1982), 91-104, Amer. Math. Soc.

[DM] D. Davis and M. Mahowald, The nonrealizability of the quotient $A//A_2$ of the Steenrod algebra, *Amer. J. Math.* **104**(1982), 1211-1216.

[GP] V. Giambalvo and D.J. Pengelley, The homology of MSpin, *Math. Proc. Camb. Phil. Soc.* **95**(1984), 427-436.

[GPR] V. Giambalvo, D.J. Pengelley and D.C. Ravenel, to appear.

[H] D. Husemoller, The structure of the Hopf algebra $H_*(BU)$ over a $Z_{(p)}$-algebra, *Amer. J. Math.* **93**(1971), 329-349.

[K] S. Kochman, An algebraic filtration of H_*BO, Proc., Northwestern Univ. Homotopy Theory Conf., *Contemporary Math.* **19**(1983), 115-143, Amer. Math. Soc.

[L] T. Lance, Steenrod and Dyer-Lashof operations on BU, *Trans. Amer. Math. Soc.* **276**(1983), 497-510.

[S] R. Stong, *Notes on cobordism theory*, Princeton University Press, 1968.

University of Connecticut, Storrs, Connecticut 06268
New Mexico State University, Las Cruces, New Mexico 88003
University of Washington, Seattle, Washington 98195

Contemporary Mathematics
Volume **58**, Part II, 1987

Sums of Squares Formulae near the Hurwitz-Radon Range

Kee Yuen Lam and *Paul Y.H. Yiu*

1. Introduction

As is well known, a sums of squares formula of type $[r, s, n]$

$$(x_1{}^2 + \cdots + x_r{}^2)(y_1{}^2 + \cdots + y_s{}^2) = f_1{}^2 + \cdots + f_n{}^2,$$

where $f_1, \ldots, f_n$ are bilinear forms with real coefficients in $x_1, \ldots, x_r$ and $y_1, \ldots, y_s$, is equivalent to each of the following:

(i) a normed bilinear map $f: \mathbf{R}^r \times \mathbf{R}^s \longrightarrow \mathbf{R}^n$ satisfying

$$|f(x, y)| = |x||y|$$

for $x \in \mathbf{R}^r, y \in \mathbf{R}^s$;

(ii) a system of $s \times n$ matrices $A_1, \ldots, A_r$ satisfying

$$A_i A_i{}^t = I_s, \quad 1 \leq i \leq r;$$

$$A_i A_j{}^t + A_j A_i{}^t = 0, \quad 1 \leq i, j \leq r, \quad i \neq j.$$

For given n and s, denote by $\rho(n, s)$ the greatest integer r for which an $[r, s, n]$ formula exists. For $s = n$, Hurwitz [5] and Radon [10] solved the matrix equations (ii) and obtained $\rho(n, n) = \rho(n) = 8a + 2^b$ for $n = 2^{4a+b}(2m + 1), 0 \leq b \leq 3$. These matrix equations, however, are much harder to handle even if n is just a few units greater than s. (See [3],for example). In this paper, we shall adopt the viewpoint (i) of normed bilinear maps and determine $\rho(s + \varepsilon, s)$ for $0 \leq \varepsilon \leq 5$ with one possibly exceptional case, by using recent results of [8],[9],[11] on the topology behind an $[r, s, n]$ formula. Specifically, for given s and ε, let $\rho = max\{\rho(s + i) : 0 \leq i \leq \varepsilon\}$ and $c = min\{j > \varepsilon : \binom{s+\varepsilon}{j} \equiv 1(mod.2)\}$.

Theorem 1. If $0 \leq \varepsilon \leq 5$, then

$$\rho(s + \varepsilon, s) = \begin{cases} \rho, & \text{if } \rho \geq 9, \\ c, & \text{if } \rho \leq 8; \end{cases}$$

except possibly for the case $\varepsilon = 5, s \equiv 27(mod.32)$.

0271-4132/87 $1.00 + $.25 per page

Remark 2. Under the conditions of Theorem 1, suppose $\rho \leq 8$ so that $\rho(s+\varepsilon, s)$ is given by c. The only cases in which c exceeds ρ are given below:

ε	s	c	ρ
2	$1(mod.4)$	3	2
4	$2(mod.8)$	6	4
4	$1,3(mod.8)$	5	4
5	$1,2(mod.8)$	6	4

2. Hurwitz-Radon matrices

For the purpose of reference and comparison, we record in this section an explicit solution of the Hurwitz-Radon equations. For details, see Yiu [11].

(i) For $1 \leq b \leq 3$, identify $\mathbf{R}^{2^b}$ as the algebras of complex numbers, quaternions and Cayley numbers respectively, with basis $e_0 = 1, e_1, \ldots, e_{2^b-1}$ satisfying

$$e_i^2 = -1, \quad 1 \leq i \leq 2^b - 1;$$

$$e_i e_j = -e_j e_i, \quad 1 \leq i,j \leq 2^b - 1, \quad i \neq j.$$

Let $E_{b,j}, 0 \leq j \leq 2^b - 1$, be the matrix corresponding to left multiplication by e_j. These, then, form a system of $\rho(2^b) = 2^b$ Hurwitz-Radon matrices of order 2^b.

(ii) Let $E = \begin{pmatrix} 0 & 1 \\ -1 & 0 \end{pmatrix}$ and $T = \begin{pmatrix} 1 & 0 \\ 0 & -1 \end{pmatrix}$. For $n = 16$, there are $\rho(16) = 9$ Hurwitz-Radon matrices given by

$$E_{4,0} = I_{16};$$
$$E_{4,i} = T \otimes E_{3,i}, \quad 1 \leq i \leq 7;$$
$$E_{4,8} = E \otimes I_8.$$

(iii) Finally, let K be the matrix $\begin{pmatrix} 0 & 1 \\ 1 & 0 \end{pmatrix} \otimes I_8$ of order 16, and let $K^{\otimes h} = K \otimes \cdots \otimes K$ (h factors). For $n = 2^{4a+b}(2m+1), 0 \leq b \leq 3$, let

$$A_0 = I_n;$$
$$A_{8h+i} = I_d \otimes (K^{\otimes h}) \otimes E_{4,i}, \qquad d = n/2^{4h+4}, \quad 0 \leq h \leq a-1, \quad 1 \leq i \leq 8;$$
$$A_{8a+j} = I_{2m+1} \otimes (K^{\otimes a}) \otimes E_{b,j}, \qquad 1 \leq j \leq 2^b - 1, \quad 1 \leq b \leq 3.$$

Then, $A_k, 0 \leq k \leq 8a + 2^b - 1$ form a system of $\rho(n) = 8a + 2^b$ Hurwitz-Radon matrices of order n.

3. Lower and upper bounds of $\rho(s+\varepsilon, s)$

For any given s and ε, we can always obtain examples of $[r, s, s+\varepsilon]$ formulae by composing normed bilinear maps of the Hurwitz-Radon type with appropriate inclusions. For $0 \le i \le \varepsilon$, the composite

$$\mathbf{R}^{\rho(s+i)} \times \mathbf{R}^{s} \hookrightarrow \mathbf{R}^{\rho(s+i)} \times \mathbf{R}^{s+i} \longrightarrow \mathbf{R}^{s+i} \hookrightarrow \mathbf{R}^{s+\varepsilon}$$

yields a $[\rho(s+i), s, s+\varepsilon]$ fomula so that $\rho(s+\varepsilon, s) \ge \rho = max\{\rho(s+i): 0 \le i \le \varepsilon\}$. On the other hand, a normed bilinear map is necessarily nonsingular. It follows from the Hopf-Stiefel condition that $\rho(s+\varepsilon, s) \le c$. (See [8],for example). Thus, we have

Proposition 3. For any s and ε,

$$\rho \le \rho(s+\varepsilon, s) \le c.$$

4. Proof of Theorem 1: the case $\rho \le 8$

If $\rho \le 8$, then $s = 16l - t$ for some l and t, $\varepsilon < t < 16$. Consequently, $c \le 8$ as well. In this case, we can obtain a $[c, s, s+\varepsilon]$ formula by restricting the multiplication table of the Cayley-Dickson algebra $\mathbf{A}_n$, $2^n \ge max(s+\varepsilon, 8)$, (cf.[2]) to the first c rows and the first s columns. See [11] for details. Thus, it follows from Proposition 3 that $\rho(s+\varepsilon, s) = c$.

This number c exceeds ρ in the cases listed in Remark 2. For example, for $s = 21$ and $\varepsilon = 2$, the nearest Hurwitz-Radon range is $[2, 22, 22]$, and the restriction process in paragraph 3 gives a $[2, 21, 23]$ formula with $\rho = 2$. However,we note that there is a formula of type $[3, 21, 23]$ "within" the multiplication of the Cayley-Dickson algebra $\mathbf{A}_5$, so that $c = 3 > \rho$. The existence of a $[3, 21, 23]$ formula can also be established without reference to $\mathbf{A}_5$, by the following considerations:

(i) there is obviously a formula of type $[3, 1, 3]$;

(ii) take a Hurwitz-Radon formula of type $[4, 20, 20]$ but restrict it to $[3, 20, 20]$;

(iii) an obvious direct sum process now gives a formula of type $[3, 1+20, 3+20]$, namely, $[3, 21, 23]$.

Indeed, all the above-mentioned $[c, s, s+\varepsilon]$ formulae,($\varepsilon \le 5$), contained within the multiplication of a Cayley-Dickson algebra can be constructed explicitly by such a direct sum process. What Theorem 1 says is that in the "near Hurwitz-Radon range" of $\varepsilon \le 5$, the process of restriction and direct sum already yields $[r, s, s+\varepsilon]$ formulae with the greatest possible r, except possibly for the case $\varepsilon = 5, s \equiv 27(mod.32)$. The story is quite different for $\varepsilon = 6$, because there is a $[10, 10, 16]$ formula (see [8] for example), which is neither a restriction of a formula of Hurwitz-Radon type, nor a direct sum of two formulae of lower dimensions.

5. Proof of Theorem 1: the cases $\varepsilon = 0, 1, 2$

The case $\varepsilon = 0$ is, of course, classical. In general,the existence of a bilinear map $\mathbf{R}^r \times \mathbf{R}^s \longrightarrow \mathbf{R}^{s+\varepsilon}$ which is normed, or even just nonsingular, will imply that, over the real projective space $\mathbf{P}^{r-1}$, there are s independent sections to the vector bundle $(s+\varepsilon)\xi = \xi \oplus \cdots \oplus \xi$ ($s+\varepsilon$ summands), where ξ denotes the Hopf line bundle. The complement of these sections would be an ε-dimensional vector bundle η such that

$$(s+\varepsilon)\xi = s\epsilon \oplus \eta. \tag{\#}$$

If $\varepsilon = 1, 2$, η is necessarily a sum of line bundles, and (#) forces $r = \rho$ in case $\varepsilon = 1$,and $r = max(3, \rho)$ in case $\varepsilon = 2$. This kind of argument is already given in Theorem 1.11 of [6].

6. Nonexistence of nonsingular bilinear maps

For $\varepsilon = 3, 4$, we again start from equation (#) and study the ε-dimensional bundle η there. This time, η may not split as a sum of line bundles, but we can use Adams' results [1] on stable vector bundles of low geometric dimensions.

For given s, let s' be the least multiple of 16 not less than s. Assume $\rho \geq 9$ so that $\rho = \rho(s')$. We quote

Theorem 4 (Adams [1]). Let $x \in \widetilde{KO}(P^\rho)$ be the stable class of the Hopf line bundle ξ_ρ over the real projective space P^ρ. The only elements in $\widetilde{KO}(P^\rho)$ that

(i) have geometric dimensions ≤ 3 are $mx, 0 \leq m \leq 3$, if $\rho \geq 13$;

(ii) can be represented by Spin(4)-bundles are 0 and $4x$;

(iii) can be represented by Spin(5)-bundles with $w_4 \neq 0$ are $4x$ if $\rho \geq 13$; $4x, -12x$ if $9 \leq \rho \leq 11$; and $4x, -12x, -60x$ if $\rho = 12$.

With this theorem, we can now have

Theorem 5. For $\varepsilon = 3, 4$, the greatest integer r for which a nonsingular bilinear map $\mathbf{R}^r \times \mathbf{R}^s \longrightarrow \mathbf{R}^{s+\varepsilon}$ exists is given by ρ if $\rho \geq 9$, and by c if $\rho \leq 8$, except possibly for

(i) $s \equiv 62, 63 (mod. 128)$ if $\varepsilon = 3$;

(ii) $s \equiv 14 (mod. 16), 61, 62 (mod. 128)$ if $\varepsilon = 4$.

Proof. If $\rho \leq 8$, this follows from paragraph 4. It remains to consider s and ε with $\rho \geq 9$. These are

(i) $\varepsilon = 3, s \equiv 13, 14, 15, 16 (mod. 16)$;

(ii) $\varepsilon = 4, s \equiv 12, 13, 14, 15, 16 (mod. 16)$.

By Proposition 3, we need only prove that there is no nonsingular bilinear map $\mathbf{R}^{1+\rho} \times \mathbf{R}^{s} \longrightarrow \mathbf{R}^{s+\varepsilon}$. All cases being similar, we shall only treat $\varepsilon = 4$ and $s \equiv 13(mod.16)$. In this case, $s' = s + 3$ and (#) becomes

$$(s'+1)\xi_\rho = (s'-3)\epsilon \oplus \eta, \tag{$*$}$$

for a 4-plane bundle η. It follows that the stable class of η is $(s'+1)x$ and that of $\varsigma = (\eta \oplus \epsilon) \otimes \xi_\rho$ is $(4 - s')x$. Now, ς is a Spin(5)-bundle with $w_4 \neq 0$. This contradicts Theorem 4 if $\rho \geq 13$ or $\rho = 10$. For $\rho = 9$,($*$) would imply 13 sections of $17\xi_9$, which contradicts Theorem 3.1 of [7]. The remaining case $\rho = 12$ corresponds to the possible exception $s \equiv 61(mod.128)$.

7. Proof of Theorem 1: the cases $\varepsilon = 3, 4$

It is not known whether nonsingular bilinear maps can actually exist in the undecided cases in Theorem 5, namely,of the types

(i) $\mathbf{R}^{13} \times \mathbf{R}^{128l+62} \longrightarrow \mathbf{R}^{128l+65}$, or even $\mathbf{R}^{13} \times \mathbf{R}^{128l+61} \longrightarrow \mathbf{R}^{128l+65}$,
(ii) $\mathbf{R}^{13} \times \mathbf{R}^{128l+63} \longrightarrow \mathbf{R}^{128l+66}$, or even $\mathbf{R}^{13} \times \mathbf{R}^{128l+63} \longrightarrow \mathbf{R}^{128l+67}$,
(iii) $\mathbf{R}^{1+\rho} \times \mathbf{R}^{16l-2} \longrightarrow \mathbf{R}^{16l+2}$, $\rho = \rho(16l)$.

However, we shall exploit properties peculiar to normed bilinear maps to show that normed bilinear maps of these types do not exist. Lam [8],[9] has shown that behind a normed bilinear map $f: \mathbf{R}^r \times \mathbf{R}^s \longrightarrow \mathbf{R}^n$, there are hidden nonsingular bilinear maps of the form $\mathbf{R}^{r+s-q} \times \mathbf{R}^q \longrightarrow \mathbf{R}^n$, and at least one of these q can be chosen in the range $r\#s \leq q \leq n$, $r\#s$ being the least integer k for which a nonsingular bilinear map $\mathbf{R}^r \times \mathbf{R}^s \longrightarrow \mathbf{R}^k$ exists. As a matter of fact, Yiu [11] proved that all hidden nonsingular bilinear maps in [8],[9] are homotopic to normed bilinear maps so that one can actually speak of hidden normed bilinear maps. With these, we have

Proposition 6. There are no normed bilinear maps of the types (i) to (iii) above.

Proof. We treat only (iii), the other cases being similar. Suppose, for a contradiction, that there is a normed bilinear map $\mathbf{R}^{1+\rho} \times \mathbf{R}^{16l-2} \longrightarrow \mathbf{R}^{16l+2}$. Observe that $(1+\rho)\#(16l-2) \geq 16l+1$, for,as in paragraph 5, the existence of a nonsingular bilinear map $\mathbf{R}^{1+\rho} \times \mathbf{R}^{16l-2} \longrightarrow \mathbf{R}^{16l}$ would imply $16l\xi_\rho = (16l-2)\epsilon \oplus \eta$ for a 2-plane bundle η which, by Stiefel-Whitney classes consideration, is necessarily 2ϵ, a contradiction. Accordingly,at least one hidden nonsingular bilinear map is of the type $\mathbf{R}^{\rho-2} \times \mathbf{R}^{16l+1} \longrightarrow \mathbf{R}^{16l+2}$ or $\mathbf{R}^{\rho-3} \times \mathbf{R}^{16l+2} \longrightarrow \mathbf{R}^{16l+2}$. Since $\rho \geq 9$, these maps do not exist by the argument in paragraph 5.

Thus, we have proved Theorem 1 for $\varepsilon = 3, 4$.

8. Proof of Theorem 1: the case $\varepsilon = 5$

This can be established by a method similar to the proof of Proposition 6. The restriction $s \not\equiv 27 (mod.32)$ is not easy to remove. It would be desirable to show that nonsingular bilinear maps of the type $\mathbf{R}^{11} \times \mathbf{R}^{16l+27} \longrightarrow \mathbf{R}^{16l+32}$ do not exist. But we have not yet managed to do so. In this regard, the work of Davis-Gitler-Mahowald [4] may be relevant.

References.

1. J.F.Adams, *Geometric dimensions of vectors bundles over RP^n*, Proc. Int. Conf. on *Prospects of Math.,Kyoto 1973*,pp.1-17, Res. Inst. Math. Sci.,Kyoto Univ., Kyoto,1974.
2. J.Adem, *Construction of some normed maps*, *Bol. Soc. Mat. Mexicana*, **20** (1975) 59-75.
3. M.A.Berger and S.Friedland, *The generalized Radon-Hurwitz numbers*; preprint.
4. D.M.Davis, S.Gitler and M.Mahowald, *The stable geometric dimension of vector bundles over real projective spaces*, Trans. *AMS*.**268** (1981) 39-61.
5. A.Hurwitz, *Über der Komposition der quadratischer Formen*, *Math. Ann.***88** (1923) 1-25.
6. K.Y.Lam, *Thesis*, Princeton University, 1966.
7. K.Y.Lam, *Sectioning vector bundles over real projective spaces* ,Quart. *J. Math. Oxford* (2)**23** (1972) 97-106.
8. K.Y.Lam, *Topological methods for studying the composition of quadratic forms*, *Canadian Math. Soc. Conf. Proc.* vol.**4** (1984) 173-192.
9. K.Y.Lam, *Some new results on composition of quadratic forms*, to appear in *Invent. Math.* 1985.
10. J.Radon, *Lineare scharen orthogonalen Matrizen*, *Abh. Math. Sem. Univ. Hamburg* **1** (1922) 1-14.
11. P.Yiu, *Thesis*, University of British Columbia, 1985.

This work is supported by the NSERC of Canada.

Department of Mathematics
University of British Columbia
Vancouver, B.C. Canada V6T 1Y4

Contemporary Mathematics
Volume **58**, Part II, 1987

Toward a Global Understanding of the Homotopy Groups of Spheres

Mark E. Mahowald
and
Douglas C. Ravenel

In this paper we will describe a point of view that has emerged as a result of research on the homotopy groups of spheres in the last decade. This philosophy is difficult to translate into theorems or even into precise conjectures, and it is certainly not apparent in the formal literature on the subject. With the exception of Theorem 10, we will not present any proofs or announcements of new results here. Rather we will collect numerous old results and current ideas and arrange them into what we hope is a suggestive picture.

1. General facts about homotopy groups

For the last 50 years one of the basic problems in algebraic topology has been the determination of the homotopy groups of spheres $\pi_{n+k}(S^n)$, i.e. the classification of continuous maps

$$S^{n+k} \to S^n$$

up to continuous deformation. The simplicity of the spaces involved lends intuitive appeal to the problem, but experience has shown that it is as hard as any in mathematics. There have been several major computational breakthroughs in the subject, namely the EHP sequence (to be described in Section 7 below), and the spectral sequences of Serre, Adams and Novikov. Each of these had lead to a large amount of new information but has also increased our appreciation of the difficulty of the problem. Much of this material is dealt with in greater depth and with numerous references in [R1].

We begin by recalling some of the basic facts about the problem. All of these groups are abelian and finitely generated. The groups $\pi_{n+k}(S^n)$ are known to vanish when $k < 0$ and when $n = 1$ and $k > 0$. The group $\pi_n(S^n)$ is isomorphic to the integers Z. These were all proved by Hurewicz around 1935. Their finite computability was established by E.H. Brown in 1959.

0271-4132/87 $1.00 + $.25 per page

The following finiteness result was proved by Serre.

Theorem 1 [S]. *The groups $\pi_{n+k}(S^n)$ for $k > 0$ are all finite with the exception of $\pi_{4n-1}(S^{2n})$, which is the direct Sigma of Z and a finite abelian group. Hence for n odd the standard map*

$$S^n \to K(Z,n)$$

induces an isomorphism in homotopy mod torsion, where $K(Z,n)$ denotes the integer Eilenberg-MacLane space. ∎

Shortly after the groups were defined Freudenthal proved that $\pi_{n+k}(S^n)$ is independent of n when $n > k+1$. These groups are said to be ***stable*** and the value of $\pi_{n+k}(S^n)$ for large n is called the ***stable k-stem*** and denoted by $\pi^{S_k}(S^0)$ or simply π^{S_k}.

The groups $\pi_{n+k}(S^n)$ with $n \leq k+1$ are called ***unstable***. We have more machinery for computing stable groups than unstable ones. One of the major themes in the subject has been the attempt to bring the more advanced technology of stable homotopy to bear on the problems of unstable homotopy theory. Two early examples of this are [A1] and [M1].

Even the stable groups are quite mysterious. The following table gives their values for small k.

k	π_k^S	k	π_k^S
0	Z	7	$Z/(240)$
1	$Z/(2)$	8	$(Z/(2))^2$
2	$Z/(2)$	9	$(Z/(2))^3$
3	$Z/(24)$	10	$Z/(6)$
4	0	11	$Z/(504)$
5	0	12	0
6	$Z/(2)$	13	$Z/(3)$

As the reader can see, these groups do not fall into any obvious pattern. However, there is a certain overall structure which we shall describe presently.

These groups form a graded ring under smash product of maps between spheres in the following way.

Given $\alpha \in \pi_j^S$ and $\beta \in \pi_k^S$, choose maps

$$f : S^{m+j} \to S^m \quad \text{and}$$
$$g : S^{n+k} \to S^n$$

for m and n sufficiently large. Then the smash product $f \wedge g$ represents the class $\alpha\beta$. It can be shown that this product is commutative up to the usual sign.

There is also a product defined in terms of composition of maps. As long as everything is in the stable range, the composition and smash products agree. However, the composition product fails to commute (even up to sign) in general. For example, if

$$f : S^3 \to S^2$$

is the Hopf map then one has

$$f \cdot 4 = 2 \cdot f$$

where "2" denotes the degree 2 map on S^2 and "4" denotes the degree 2 map on S^4.

This noncommutativity is very important in unstable homotopy theory.

In [N] Nishida proved that every positive dimensional element in this ring is nilpotent, i.e. for each $\alpha \in \pi_k^S$ with $k > 0$, there is an exponent m such that $\alpha^m = 0$. A vast generalization of this nilpotence result conjectured in [R2] has been proved recently by Devinatz, Hopkins and Smith and will be commented on below (see Theorem 6). Nishida's Theorem indicates that the multiplicative structure of this ring is of very limited use. One should not attempt to describe it in terms of generators and relations.

2. Periodic families

A construction which has proved quite useful is the following. Suppose we have a finite complex Y and a map

$$v : \Sigma^d Y \to Y$$

such that *all* iterates of the form

$$v^i = v \cdot \Sigma^d v \cdots \Sigma^{(i-1)d} v : \Sigma^{id} Y \to Y$$

are essential. It is convenient for technical reasons to require that the complex Y and the self-map v are both double suspensions. A map of this sort all of whose iterates are essential (i.e. not homotopic to the constant map) is said to be *periodic*.

The set of homotopy classes of maps from $\Sigma^k Y$ to a space X is an abelian group, which we will denote by $\pi_k(X;Y)$. (We reserve the right to change the index k when convenient.) Given an element $\alpha \in \pi_k(X;Y)$ represented by a map

$$f : \Sigma^k Y \to X$$

we can define $v^i\alpha \in \pi_{k+id}(X;Y)$ to be the class represented by the composite $f \cdot \Sigma^k v^i$.

This makes the graded group $\pi_*(X;Y)$ a module over the ring $Z[v]$. If the finite complex Y is such that the identity map has order q in the group $[Y,Y]$ then $\pi_*(X;Y)$ is a module over the ring $Z/(q)[v]$. In any case one can tensor with $Z[v, v^{-1}]$ and ask the following question.

What is $v^{-1}\pi_*(X : Y)$?

An element in $\pi_*(X;Y)$ has a nontrivial image in this group iff it is not annihilated by any power of v. Such elements are said to be *v-periodic*. Elements which are annihilated by some power of v are said to be *v-torsion*. In Section 5 we will introduce BP-theory and explain how it provides us with some powerful methods for computing or at least estimating this group. A more delicate question is that of the image of $\pi_*(X;Y)$ in it. An answer to this question will lead to some useful information about $\pi_*(X)$ itself.

A *v-periodic family* in $\pi_*(X;Y)$ is a set of elements of the form

$$\{x, vx, v^2x, \cdots\}$$

such that v^ix is nontrivial for all $i > 0$.

The finiteness theorem of Serre mentioned above can be regarded as a result of this type. Let $X = S^n, Y = S^1$ and let v be the degree p map for p a prime. Then we are asking for the structure of $p^{-1}\pi_*(S^n)$. If we localize everything in sight at the prime p we get rid of torsion prime to p, so the image of $\pi_*(S^n)$ is its torsion free quotient. For n odd Serre's theorem says that the standard map from S^n to $K(Z,n)$ induces a homotopy equivalence modulo torsion, i.e. the map is a rational homotopy equivalence.

Notice that Nishida's theorem tells us that if we take our finite complex Y to be a sphere of any dimension then the degree p map is essentially the only possible choice for the self map v. Any map coming from a positive stem would have to be nilpotent and therefore would not be suitable. The result of Devinatz-Hopkins-Smith (see Theorem 6 below), which generalizes Nishida's theorem tells us that the same is true whenever $H_*(Y)$ is torsion free. It turns out that using any such Y will give us essentially the same information about $\pi_*(X)$, namely its torsion free quotient.

3. The stable image of J as a periodic family

In view of the remarks in the previous paragraph we should consider a finite complex Y with some torsion in its homology. Let $M(p)$ denote a mod (p) Moore space. We do not care about the dimension of its bottom cell, as long as it is not too small. We will use the same notation for the mod (p) Moore ***spectrum*** with its bottom cell in dimension 0. This is now our finite complex Y. The group $\pi_*(X;Y)$ is called the ***mod*** (p) ***homotopy*** of X. The self map v is provided by the following result of Adams.

Lemma 2 [A2]**.** *Let q be $2p-2$ if p is an odd prime and 8 if $p=2$. Then there is a map*

$$v : \Sigma^q M(p) \to M(p)$$

inducing an isomorphism in K-theory. Hence all iterates of this v are nontrivial.

The stable composite

$$S^q \to \Sigma^q M(p) \xrightarrow{v} M(p) \to S^1$$

is α_1 (the first element of order p in π^{S_*}) for p odd and 8σ (where σ is the generator of π^{S_7}) for $p=2$. The first and third maps here are respectively the inclusion of the bottom cell in $M(p)$ and the pinch map from $M(p)$ to its top cell.

If X is the sphere spectrum then $\pi_*(X;M(p))$ is the set of stable maps from $M(p)$ to a sphere. By S-duality this is isomorphic to $\pi^{S_*}(M(p))$ (up to reindexing) with v acting by composition on the right instead of the left.

More generally for a finite spectrum Y we have a similar isomorphism

$$\pi_*(X;Y) = \pi_*(X \wedge DY)$$

where DY is the Spanier-Whitehead dual of Y, i.e. the complement of the finite complex Y embedded in a suitable sphere. In most cases of interest (e.g. $Y = M(p)$), Y is self-dual, i.e. DY is some suspension of Y.

For $p=2$ the v-torsion free quotient of

$$\pi_*(S^0; M(p))$$

was first determined in [M3]. The following odd primary analog was proved by Miller.

Theorem 3 [Mi]**.** *For an odd prime p, $\pi^{S_*}(M(p))$ mod v-torsion is generated by two elements represented by the stable composites*

$$\begin{array}{ccccc} S^q & \longrightarrow & \Sigma^q M(p) & \xrightarrow{v} & M(p) \quad \text{and} \\ S^{q-1} & \xrightarrow{\alpha_1} & S^0 & \longrightarrow & M(p). \end{array}$$

The result for $p = 2$ is considerably more complicated. There are ten generators instead of two and the group is a module over $Z/(4)[v]$ instead of $Z/(2)[v]$. The details need not concern us here.

What does this tell us about the stable homotopy groups of spheres? There is a long exact sequence relating ordinary and mod (p) stable homotopy groups. In it the mod (p) groups in Miller's theorem correspond to the p-component of the image of the stable J-homomorphism, which we will now describe.

There is a homomorphism

$$J : \pi_k(SO(n)) \to \pi_{n+k}(S^n)$$

defined originally by Hopf and Whitehead. Here $SO(n)$ denotes the group of $n \times n$ orthogonal matrices with determinant one.

Letting n go to ∞ gives a homomorphism

$$J : \pi_k(SO) \to \pi_k^S(S^0)$$

where SO denotes the stable orthogonal group. Its homotopy groups were determined by Bott, who showed that

$$\begin{aligned} \pi_k(SO) &= Z \text{ if } k \equiv 3 \bmod (4), \\ &= Z/(2) \text{ if } k \equiv 0 \text{ or } 1 \bmod (8) \quad \text{and} \\ &= 0 \text{ otherwise.} \end{aligned}$$

Adams showed in [A2] that J is monomorphic on the 2-torsion and that its image on each free Sigmamand is a cyclic group whose order is a certain arithmetic function of the dimension k. (Actually his work left an ambiguous factor of two which depended on the Adams conjecture.) In particular, the order of this cyclic group is divisible by an odd prime p iff $k \equiv -1 \bmod (2p-2)$.

4. Some unstable results

An unstable analog of Theorem 3 would describe the v-torsion free quotient of $\pi_*(S^n; M(p))$. Such a result for n odd and $p = 2$ was obtained in [M2]. For technical reasons it was necessary to replace $M(2)$ by the complex $Y = M(2) \wedge CP^2$ in order to get the cleanest possible description of the actual groups, but this should be regarded as a minor detail.

Recall that Serre's theorem gave a map to an infinite loop space, namely $K(Z,n)$, which induced an isomorphism modulo p-torsion. Ideally one would like to have a similar map to an infinite loop space inducing an isomorphism in mod (p) homotopy modulo v-torsion, where v is as in Adams' lemma above. The best we can do is the following.

Theorem 4 [M2]. *Let $n = 2m+1$ and $p = 2$. There is a map*

$$f : \Omega_0^{2m+1} S^{2m+1} \to \Omega^\infty RP^{2m} \wedge J$$

inducing an isomorphism in $\pi_*(\ ; M(2))$ modulo v-torsion. Here the source is the degree 0 component of the indicated loop space and the target is the 0th space in the Ω-spectrum $RP^{2m} \wedge J$ where RP^{2m} is $2m$-dimensional real projective space and J is a certain spectrum which will be described below. ∎

This result is useful because the homotopy groups of the target can be computed explicitly. For example the mod (2) homotopy of J is v-torsion free. This fact strengthens the analogy with Serre's theorem, which gave a map from S^{2m+1} to an infinite loop space with torsion free homotopy.

The theorem does not assert that the map induces a surjection in homotopy or in mod (2) homotopy, but that the mod (2) kernel is precisely the v-torsion subgroup and that the cokernel is all v-torsion. In [M2] the cokernel is described up to an ambiguity related to the Kervaire invariant problem.

It is quite likely that there is an analogous result for every odd prime with RP^{2m} replaced by the $2m(p-1)$-skeleton of $B\Sigma_p$, the classifying space for the symmetric group on p letters.

Conjecture 5. The map f in the theorem above induces an isomorphism in K-theory.

The K-theory of the target has been computed by Miller and Snaith, but we do not know how to compute the K-theory of the source.

Now we will describe the spectrum J. It is so named because its homotopy is nearly identical to the image of the J-homomorphism. There is a different version of this spectrum for each prime p. We will use the same notation for each. We will describe the odd primary version first.

Let bu be the spectrum for connective complex K-theory. This spectrum can be built out of the various connected covers of BU, the classifying space for stable complex vector bundles. J is the fiber of a certain map from $bu_{(p)}$ (the localization of bu at the prime p) to $\Sigma^q bu_{(p)}$. This map induces an isomorphism in homotopy in dimensions not divisible by

$2p-2$. The homotopy of J in positive dimensions is isomorphic to the p-component of the image of the stable J-homomorphism.

For $p=2$ the definition of J is slightly different. It is the fiber of a map from $bo_{(2)}$ to $\Sigma^4 bsp_{(2)}$. Here bo and bsp are the spectra obtained from the $(8k-1)$-connected covers of the spaces BO and BSp, the classifying spaces for stable real and symplectic vector bundles respectively. Bott periodicity tells us that $\Omega^4 BSp = Z \times BO$ and $\Omega^4 BO = Z \times BSp$.

The homotopy of J in positive dimensions is *not* isomorphic to the image of J. The latter maps monomorphically to the former, leaving a cokernel of $Z/(2)$ in dimensions congruent to 0 and 1 mod (8). These groups correspond to the homotopy elements μ_{8k} and μ_{8k+1} constructed in [A2].

For each prime p the stable map

$$S^0(p) \to J$$

induces an isomorphism in K-theory, a surjection in homotopy and an isomorphisn in $v^{-1}\pi_*(\ ; M(p))$.

5. Enter BP-theory

Recall that our basic setup is the following. We have a finite complex Y and a map v to Y from some suspension of Y such that all iterates of v are nontrivial. Then we try to compute $v^{-1}\pi_*(X; Y)$ for our favorite space X.

We have seen three examples of reslts of this sort, namely Theorems 1, 3 and 4. In Theorems 1 and 4, $X = S^{2n+1}$. In the former case $Y = S^1$ and v is the map of degree p. In Theorems 3 and 4, Y is a mod (p) Moore space or spectrum and the self-map v is that provided by Lemma 2.

Before proceeding to the main theorem in each case one must prove that all of the iterates of v are nontrivial. In the case of the degree p map in Theorem 1 this can be done with ordinary integer homology. Since $H_1(S^1) = Z$, we know that the map of degree p^i (the ith iterate of v) is nonzero for all i. In other words all iterates of the maps are essential because the map induces an isomorphism in rational homology.

The self-maps of Lemma 2 (used in Theorems 3 and 4) have nontrivial iterates because they induce isomorphisms in K-theory. K-theory could also be used to show that all iterates of the degree p map are nontrivial. Ordinary homology would *not* suffice to prove that the maps of Lemma 2 are nontrivial.

BP-theory, like K-theory, is a generalized homology theory strong enough to show that these two self-maps are periodic, i.e. that all of their iterates are essential. We will explain how it applies in these two cases.

There is a different version of BP-theory for each prime p. (If one wants to work globally one can use MU-theory, which is the same thing as complex cobordism.) For any space X, $BP_*(X)$, the reduced BP-homology of X, is a graded module over the coefficient ring

$$BP_* = Z_{(p)}[v_1, v_2, \cdots]$$

where the dimension of v_i is $2p^i - 2$ and $Z_{(p)}$ denotes the integers localized at the prime p.

The degree p map on S^n induces multiplication by p in

$$BP_*(S^n) = \Sigma^n BP_*.$$

For the self-maps in Lemma 2 we have $BP_*(M(p)) = BP_*/(p)$ and v induces multiplication by v_1 for p odd and by V_1^4 for $p = 2$. In the latter case there is no self map of $M(2)$ inducing multiplication by simply v_1. The 4-cell complex $Y = M(2) \wedge CP^2$ used in [M2] was chosen precisely becaue it does admit such a self-map.

Since the Adams self-maps of Lemma 2 induce multiplication by v_1 or some power of it, the homotopy elements derived from Theorems 3 and 4 are said to be *v_1-periodic.*

What other sorts of finite complexes with periodic self-maps might we have?

Several conjectures conerning this question were made in [R2] and a lot of progress has been made on them in the past three years. The most striking result, proved very recently by Devinatz, Hopkins and J. Smith, says that BP-theory is strong enough to detect any such self-map.

Nilpotence Theorem 6 [DHS]. *Let Y be a p-local finite complex and let*

$$v : \Sigma^d Y \to Y$$

be a map inducing the trivial homomorphism is BP-homology. Then some iterate of v is null homotopic. ∎

Note that Nishida's theorem is a special case of this, since it is easy to show that any self-map of a sphere in a positive stem induces the trivial map in BP-homology.

It is known that for any finite complex $Y, BP_*(Y)$ is finitely presented as a BP_*-module. From this it follows easily that any periodic endomorphism of $BP_*(Y)$ has an iterate which is an idempotent composed with multiplication by some element v in BP_*.

For the sake of simplicity, asSigmae that this idempotent is the identity. Then the internal properties of BP-theory tell us essentially that this v must be a power of one of the polynomial generators v_n.

Moreover n is uniquely determined by the module $BP_*(Y)$ in the following way. It is the smallest n such that $v_n^{-1}BP_*(Y)$ is nonzero, i.e. such that $BP_*(Y)$ contains elements not annihilated by any power of v_n. It is known [JY] that if $v_n^{-1}BP_*(Y)$ is nonzero then so is $v_m^{-1}BP_*(Y)$ for all $m > n$. In this case we say that the finite complex Y has ***type n***. Until 1983 it was not even known that such finite complexes exist for all n [Mit]. A conjecture in [R2] that remains open says that every finite complex admits a periodic self-map.

$Y = M(p)$ has type 1. If Y has type n then we can denote $v^{-1}\pi_*(X;Y)$ by $v_n^{-1}\pi_*(X;Y)$ and refer to the v-periodic and v-torsion elements as v_n-periodic and v_n-torsion elements respectively. We conjecture that the information $v^{-1}\pi_*(X;Y)$ gives about X depends only on n. More precisely,

Conjecture 7. Let Y and Y' be two p-local finite complexes of type n and let f be a map from X_1 to X_2. Then the following are equivalent:

(i) f induces an isomorphism in $v_n^{-1}\pi_*(\ ;Y)$,

(ii) f induces an isomorphism in $v_n^{-1}\pi_*(\ ;Y')$, and

(iii) f induces an isomorphism in $v_n^{-1}BP_*(\ )$. ∎

6. The chromatic filtration

Now we will describe a way to codify all the information given by the groups $v_n^{-1}\pi_*(X;Y)$ for all n in the stable case, i.e. when X and Y are spectra.

We first need to introduce Bousfield localization [B]. Let E_* be a generalized homology theory. A spectrum Y is *E-local* if for every E-acyclic spectrum X (i.e. a spectrum satisfying $E_*(X) = 0$), $[S, Y] = 0$, i.e. there are no essential maps from X to Y.

An *E-localization* of a spectrum W is a map to an E-local spectrum W' which is an E_*-equivalence, i.e. a map inducing an isomorphism in E-homology. It is easy to show that if such a localization exists it is unique. If both E and W are connective then the E-localization of W simply an arithmetic localization or completion.

However, if either of them fail to be connective then the existence of the localization is far from obvious and its properties are hard to predict. For example, the K-theoretic localization of the sphere has nontrivial homotopy groups in arbitrarily large negative dimensions and its π_{-2} is Q/Z. The localization of its (-1)-connected cover at an odd prime is the spectrum J discussed earlier. More details can be found in [R2].

In [B] Bousfield proved that such localizations always exist. Let L_nX denote the localization of X with respect to $v_n^{-1}BP$. L_0X is the rational localization of X; by convention $v_0 = p$. L_1X is the localization with respect to p-local K-theory. A $v_n^{-1}BP$-local spectrum is also $v_{n+1}^{-1}BP$ so there are natural maps

$$L_0X \leftarrow L_1X \leftarrow \cdots$$

We say that X is ***harmonic*** if this inverse system converges to X. In [R2] it is shown that all p-local finite spectra are harmonic.

If Y is a finite complex of type n then one of the conjectures of [R2] implies that

$$v_n^{-1}\pi_*(X;Y) = \pi_*(L_nX;Y).$$

In other words L_nX captures all the v_n-periodic information about X.

The inverse system above leads to a decreasing filtration of $\pi_*(X)$, defined by setting $F^n\pi_*(X)$ equal to the kernel of the homomorphism induced by the localization map from X to L_nX, i.e. to the v_n-torsion. This is the ***chromatic filtration*** of $\pi_*(X)$. An algebraic analog of it, the chromatic spectral sequence, is studied extensively in [MRW] and in Chapter 5 of [R1].

Let M_nX denote the fiber of the map from L_nX to $L_{n-1}X$. Its homotopy is all v_{n-1}-torsion and all v_n-periodic. We know [R3] how to compute $BP_*(M_nX)$. Its Adams-Novikov spectral sequence is conjectured but unfortunately still not known to converge to $\pi_*(M_nX)$. It E_2-term is closely rrelated to the continuous mod (p) cohomology of a certain pro-p-group with interesting arithmetic properties.

This connection was discovered over a decade ago by Morava [Mo]. Again we refer the reader to [MRW] or [R1] for more details. It means that this E_2-term is in some sense finitely computable, which is certainly not the case for most E_2-terms associated with finite complexes. For $n = 1$ and $X = S^0$ this computation is done in complete detail in [MRW] and in Chapter 5 of [R1].

Admittedly there are infinitely many values of n required for a complete understanding of the homotopy groups of spheres, but each of these finite calculations yields an infinite amount of information.

With Theorem 4 and related stable results we have a complete picture of the v_1-periodic homotopy groups (both stable and unstable) of spheres, at least for $p = 2$.

We only have partial results for the v_2-periodic picture, which is considerably more complicated. There is no known v_2-periodic analog of either the J-homomorphism or the spectrum J.

It is also clear that things are simpler for larger primes. For $p > 3$ let $V(1)$ denote the cofiber of the Adams self-map of Lemma 2. It is the simplest complex of type 2. The conjectures of [R2] imply that

$$v_2^{-1}\pi_*(S^0;V(1))$$

has precisely 12 generators. The image of $\pi_*(S^0;V(1))$ in this group is still unknown.

7. The EHP sequence

The EHP sequence is the fundamental tool for understanding unstable homotopy groups of spheres. A more thorough exposition can be found in the last section of Chapter 1 of [R1]. For simplicity we limit our remarks to the prime 2. Most of what is said below has an analog at any odd prime, but the statements are more complicated.

There are fibrations

$$S^n \xrightarrow{E} \Omega S^{n+1} \xrightarrow{H} \Omega S^{2n+1}$$

for all positive n. The map E stands for suspension (Einhängung) and H for Hopf invariant. The associated long exact sequence of homotopy groups is the EHP sequence.

Collectively (for all n) these long exact sequences constitute an exact couple which leads to what is called the EHP spectral sequence. Many classical results about unstable homotopy groups can be translated into statements about this spectral sequence. Such results include the Freudenthal Suspension Theorem (see Section 1 above) and the Adams Vector Field Theorem of [A1]. Theorem 4 has implications concerning certain patterns in the EHP spectral sequence, specifically the behavior of elements whose Hopf invariant lies in the image of J (hence the title of [M2]).

The following result of Snaith, enhanced by an observation of Kuhn, relates unstable groups appearing in the EHP spectral sequence to stable groups.

Theorem 8 [Sn] [K]. *There are maps*

$$f : \Omega^{n+1} S^{n+1} \to QRP^n$$

(where QX denotes $\Omega^\infty\Sigma^\infty X$) such that the following diagram commutes up to homotopy.

$$\begin{array}{ccc} \Omega^n S^n & \xrightarrow{f} & QRP^{n-1} \\ \downarrow & & \downarrow \\ \Omega^{n+1} S^{n+1} & \xrightarrow{f} & QRP^n \\ \downarrow & & \downarrow \\ \Omega^{n+1} S^{2n+1} & \longrightarrow & QS^n \end{array}$$

The vertical maps on the left are $\Omega^n E$ and $\Omega^n H$ while the ones on the right are the images under the functor Q on the evident cofiber sequence. ∎

Using this result one can easily produce a map from the EHP spectral sequence to the Atiyah-Hirzebruch spectral sequence for the stable homotopy of RP^∞, in which all groups in sight are stable.

There are versions of the EHP spectral sequence converging to the stable homotopy groups of spheres and to the unstable homotopy groups of any given sphere. The E_1-term also consists of unstable homotopy groups, so it lends itself very well to inductive calculation. The input at each stem consists of the output from lower stems. In principle all one needs to start the inductive process is knowledge of the homotopy groups of S^1, which are easily determined.

The entire EHP apparatus can be adapted to the computation of $\pi_*(S^n; Y)$. The following result shows that each element in a v_n-periodic family (with a finite number of exceptions for any given family) has a v_n-periodic Hopf invariant, and that all but finitely many elements in each such family originate on the same sphere. More precisely we have

Theorem 9. *Let Y be a finite complex of type n with a self-map v inducing multiplication by some power of v_n in its BP-homology. If*

$$x \in \pi_*(S^m; Y)$$

is v-periodic, then there is a $k \leq m$ such that some v-multiple of x is the iterated suspension of

$$x' \in \pi_* S^k; Y)$$

and the Hopf invariant

$$H(x') \in \pi_*(S^{2k-1}; Y)$$

is v-periodic. Hence all higher v-multiples of x are born on S^k.

Proof. Suppose $H(x')$ is not v-periodic. Then by definition some power of v annihilates it. Thus the corresponding v-multiple of x' desuspends further since the Hopf invariant is the obstruction to desuspension. This process must stop after a finite number of steps because we cannot desuspend below the 1-sphere. Thus some v-multiple of x desuspends to an element with a v-periodic Hopf invariant as claimed, and all higher v-multiples of x are born on S^k. ∎

Very early work with the EHP sequence showed that the p-torsion subgroups of the homotopy of each finite sphere has a bounded exponent. The best possible result of this sort for odd primes was proved by Cohen-Moore-Neisendorfer [CMN]. Experimental evidence suggests that there should be similar bounds on the v_n-torsion in suitable groups for finite spheres.

8. James periodicity and the root invariant

Theorem 8 shows a connection between the EHP sequence and the homotopy groups of stunted projective spaces. For $m \le n$ let RP_m^n denote RP^n/RP^{m-1}, i.e. the stunted projective space with bottom cell in dimension m and top cell in dimension n. There is an equivalence

$$\Sigma^K RP_m^n \to RP_{m+K}^{n+K}$$

if K is a sufficiently large (depending on $n - m$) power of 2. This leads to a certain regularity in the EHP sequence called *James periodicity.* It is *not* related to the v_n-periodicity discussed elsewhere in this paper.

It is possible to define spectra RP_m^n and RP_m^∞ (denoted simply by RP_m) for arbitrary (e.g. negative) integers m. There are maps

$$RP_m \leftarrow RP_{m-1} \leftarrow \cdots$$

obtained by pinching out the bottom cell in each case.

Theorem 10 Lin [L]. *The homotopy inverse limit of the RP_m (denoted by $RP_{-\infty}$) is homotopy equivalent to the 2-adic completion of S^{-1} in such a way that the composite map*

$$S^{-1} \to RP_{-\infty} \to RP_{-1}$$

is homotopic tot he inclusion of the bottom cell. ∎

Theorem 11 Kahn-Priddy [KP]. *The inclusion of the bottom cell in RP_{-1} induces the trivial homomorphism in all homotopy groups except π_{-1}, in which the image is $Z/(2)$.* ∎

Theorem 10 means that the Atiyah-Hirzebruch spectral sequence for the stable homotopy of $RP_{-\infty}$ must converge to that of S^{-1}. This has certain implications for the EHP sequence. In particular it allows us to define a mutation of the Hopf invariant which we call the root invariant.

Suppose $\alpha \in \pi_n^S$ is represented by a map

$$S^{n-1} \to S^{-1}.$$

If n is positive or $n = 0$ and α is divisible by two then the composite

$$S^{n-1} \to S^{-1} \to RP_{-1}$$

is null homotopic by Theorem 11.

On the other hand Theorem 10 guarantees that for some m the composite

$$S^{n-1} \to S^{-1} \to RP_{-m}$$

is essential.

Taking the smallest such m we get a factorization

$$S^{n-1} \to S^{-m} \to RP_{-m}$$

defined modulo a certain indeterminacy.

The resulting coset in

$$\pi_{n-1}(S^{-m}) = \pi_{m+n-1}(S^0)$$

is the ***root invariant*** $R(\alpha)$. Notice that we started in the n-stem and ended up in the $(n+m-1)$-stem, where the number m depends on α as well as on n. It is known that in general $m \geq 2n+1$. Many root invariants have been calculated. For example we have

$$\begin{aligned} \dim R(2^i) &= 2i - 1 \text{ if } i \equiv 0 \text{ or } 1 \bmod (4) \\ &= 2i - 2 \text{ if } i \equiv 2 \bmod (4) \\ &= 2i - 3 \text{ if } i \equiv 3 \bmod (4) \end{aligned}$$

and $R(2^i)$ contains the element of order two in the image of J if $i \equiv 0$ or $3 \bmod (4)$. It is conjectured that $R(\theta_j)$ contains θ_{j+1} if both exist, where θ_j is the Kervaire invariant element.

The definition of the root invariant can be adapted to homotopy with coefficients in Y. Experimental evidence suggests that the root invariant of a v_n-periodic element is v_{n+1}-periodic. More precisely we have

Conjecture 12. If Y is of type n and has a v_n-periodic self-map v

$$\alpha \in \pi_*(S^{-1}; Y)$$

is v-periodic then the coset $R(\alpha)$ consists entirely of v_n-torsion elements. Let

$$w : \Sigma^d Y \to Y$$

be a power of v which annihilates every element in $R(\alpha)$ and let Z be its cofiber. Thus Z has type $n+1$ and each element in $R(\alpha)$ extends to a map from Z to a suitable sphere. At least one of these maps is v_{n+1}-periodic. ■

REFERENCES

[A1] J.F. Adams, Vector fields on spheres, *Ann. of Math.*, **75** (1962), 603-632.

[A2] J.F. Adams, The groups $J(X)$, IV, *Topology*, **5** (1966), 21-71.

[B] A.K. Bousfield, Localization of spaces with respect to homology, *Topology*, **18** (1979) 257-281.

[CMN] F.R. Cohen, J.C. Moore and J. Neisendorfer, The double suspension and exponents of the homotopy groups of spheres, *Ann. of Math.*, **110** (1979), 549-565.

[DHS] E. Devinatz, M. Hopkins and J. Smith, Nilpotence in stable homotopy theory, to appear.

[JY] D.C. Johnson and Z. Yosimura, Torsion in Brown-Peterson homology and Hurewicz homomorphisms, *Osaka J. Math.*, **17** (1980), 117-136.

[KP] D.S. Kahn and S.B. Priddy, The transfer and stable homotopy, *Math. Proc. Camb. Phil. Soc.*, **83** (1978), 103-111.

[K] N. Kuhn, The geometry of the James-Hopf maps, *Pac. J. Math.*, **102** (1982), 397-412.

[L] W.H. Lin, On conjectures of Mahowald, Segal and Sullivan, *Math. Proc. Camb. Phil. Soc.*, **87** (1980), 449-458.

[M1] M.E. Mahowald, The metastable homotopy of S^n, Memoirs Amer. Math. Soc. No. 72 1967.

[M2] M.E. Mahowald, The image of J in the EHP sequence, *Ann. of Math.*, **116** (1982), 65-112.

[M3] M.E. Mahowald, *bo*-resolutions, *Pacific J. Math.*, **192** (1981), 365-383.

[Mi] H.R. Miller, On relations between Adams spectral sequences, with an application to the stable homotopy of a Moore spaces, *J. Pure Appl. Alg.*, **20** (1981), 287-312.

[MRW] H.R. Miller, D.C. Ravenel and W.S. Wilson, Periodic phenomena in the Adams-Novikov spectral sequence, *Ann. of Math.*, **106** (1977), 469-516.

[Mit] S.A. Mitchell, Finite complexes with A_n-free cohomology, to appear in *Topology*.

[Mo] J. Morava, Structure theorems for cobordism comodules, to appear somewhere.

[N] G. Nishida, The nilpotency of elements in the stable homotopy groups of spheres, *J. Math. Soc. Japan*, **25** (1973), 707-732.

[R1] D.C. Ravenel, *Complex cobordism and stable homotopy groups of spheres*, Academic Press, New York, 1986.

[R2] D.C. Ravenel, Localization with respect to certain periodic homology theories, *Amer. J. Math.*, **106** (1984), 351-414.

[R3] D.C. Ravenel, The geometric realization of the chromatic resolution, Proc. of Conference in Honor of J.C. Moore, to appear.

[S] J.-P. Serre, Groupes d'homotopie et classes de groupes abelien, *Ann. of Math.*, **58** (1953), 258-294.

[Sn] V.P. Snaith, Stable decomposition of $\Omega^n \Sigma^n X$, *J. London Math. Soc.*, **7** (1974), 577-583.

Northwestern University, Evanston, IL 60201
University of Washington, Seattle, WA 98195
Both authors partially supported by the N.S.F.

Contemporary Mathematics
Volume **58**, Part II, 1987

SURGERY WITH FINITE FUNDAMENTAL GROUP

R. J. Milgram

Surgery leads to exact sequences

0.1
$$\cdots \xrightarrow{\partial} L^h_{n+1}(Z(\pi_1(M^n))) \longrightarrow HT(M^n) \longrightarrow$$
$$[M^n, G/\{TOP \text{ or } PL\}] \xrightarrow{\partial} L^h_n(Z(\pi_1(M^n))).$$

where $HT(M^n)$ is the set of h-cobordism classes of topological or PL manifolds homotopy equivalent to M^n. When $\pi_1(M^n)$ is finite the map ∂ is determined explicitly by three maps of the bordism of the 2-Sylow subgroup of π_1 to $L^h_*(Z(\pi_1))$ determined by evaluating the obstructions for problems of the form

0.2
$$id\times\sigma\colon M^n\times N^j \longrightarrow M^n\times S^j$$

where S^j is the sphere if j is even and the twisted sphere $S^1\times_T S^{j-1}$ otherwise. (T is orientation reversing on S^{j-1} and rotation through 180 degrees on S^1.) N is the index 8 Milnor manifold when j is divisible by 4, and the Kervaire manifold or the twisted Kervaire manifold otherwise. In the first case Taylor-Williams have announced that the simply connected index is the only obstruction for 0.2. and in this note we announce the results for the Kervaire case. In [3] we complete the analysis of ∂ by obtaining the answer for the twisted Kervaire case.

These results complete the determination of the sets $HT(M^n)$ when $\pi_1(M^n)$ is finite. One of the corollaries (Corollary D) is the oozing conjecture, but more generally this completes surgery classification theory for homotopy triangulations whenever the fundamental group is finite. Details appear in [3], [5], and [6].

THE RESULTS

The key is to factor Ranicki's algebraic product pairing

0.3 $$L_h^*(Z(\pi))\otimes L_j^h(Z(\pi'))\longrightarrow L_{*+j}^h(Z(\pi\times\pi'))$$

through new L-groups which are better controlled than both the domains and ranges above. We have, based on an idea of Clauwens [1],

Theorem A:

a) *When* N *is the Kervaire manifold the obstruction map* 0.3 *factors as:*

$$L_h^m(Z(\pi))\longrightarrow L_m^h(Z(\zeta_3)\pi)\longrightarrow L_{m+2}^h(Z(\zeta))$$

b) *When* N *is the Milnor manifold the obstruction map factors as:*

$$L_h^m(Z(\pi))\longrightarrow L_m^h(Z(\zeta_{15})\pi)\longrightarrow L_m^h(Z(\pi))$$

c) *When* N *is the twisted Kervaire manifold the obstruction map factors as:*

$$L_h^m(Z(\pi),\ Z/2)\longrightarrow L_m^h(Z(\tfrac{1+\sqrt{5}}{2})\pi,\ Z/2)\longrightarrow L_{m-1}^h(Z(\pi),\ Z/2).$$

In each of the intermediate groups above the involution on the coefficient domain is non-trivial (in (a), (b), complex conjugation, and in (c) the non-trivial Galois automorphism).

FOR SIMPLICITY THROUGHOUT THE REMAINDER OF THIS NOTE WE CONCENTRATE ON THE CASE OF THE KERVAIRE PROBLEM.

Theorem B: *Let* π *be a finite* 2-*group then the set of restriction followed by quotient maps*

$$L_*^h(Z(\zeta_3)\pi)\longrightarrow L_*^h(Z(\zeta_3)H)$$

where H *runs over the set of dihedral groups* $D(2^i,2)$, *quaternion groups* $Q(2^i)$, *and all the* $Z/2$, $Z/2\times Z/2$ *quotients of* π^{ab}, *detect all the possible classes which can occur as obstructions for* 0.2.

This reduces the calculation to that for the quaternion and dihedral groups, and the groups $Z/2$, $Z/2\times Z/2$.

In the next theorems we assume M^n is oriented. Let V denote the total Wu class of M, and $f\colon M\longrightarrow B_\pi$ the classifying map for the universal cover.

Theorem C:

a) *When the dimension of* M *is* 4i *the image is the index of* M.

b) *When the dimension of* M *is* 4i+1, *then the map* μ *is non-trivial if and only if there is an element* $e \in H^1(B_\pi; Z/2)$ *so that* $\langle V^2 f^*(e), [M]\rangle$ *is non-zero.*

c) *When the dimension of* M *is* 4i+2 *then* μ *is detected by projection onto a group* $Z/2\times Z/2$ *or by restriction to a subgroup, and then projection to a dihedral group. In the case of both the dihedral group and* $Z/2\times Z/2$ *there is a class*

$$w \in H^2(B_\pi; Z/2)$$

which is not in the image of Sq^1 *and the obstruction is given by* $\langle V^2 f^*(w), [M]\rangle$.

d) *If the dimension of* M *is* 4i+3, *then the map is detected on restriction and projection onto quaternion subquotients where the formula is* $\langle V^2 f^*(b_3), [M]\rangle$, *and* b *is the non-zero class in* $H^3(B_{Q(2^i)}; Z/2) = Z/2$.

Note that in each case the deviation from being a pure characteristic class formula is a class $f^*(c)$ where dimension (c) < 3. This is the content of the oozing conjecture for oriented manifolds

Corollary D: *The (codimension 3) oozing conjecture is true for oriented manifolds with finite fundamental group. That is to say, if* π *is a finite group, then the formula for the product with Kervaire obstruction above has the form* $\langle V^2 f^*(\kappa_*), [M]\rangle$ *where* $\kappa = 1 + \kappa_1 + \kappa_2 + \kappa_3$, *with the*

$$\kappa_i \in H^i(B_\pi; Z/2)$$

whenever M *is oriented and* $\pi_1(M) = \pi$.

Of course, theorems B, C make the determination of the κ's explicit for any finite group π.

This research was partially supported by the N.S.F., and evolved primarily during visits at The University of Edinburgh, Northwestern University, McMaster University, and UCSD.

Section 1. The L-groups when π is a finite 2-group.

We need some notation. Set

$$
\begin{aligned}
w_i &= \zeta_{2^i} \\
\lambda_i &= w_i + (w_i)^{-1} \qquad (1.1)\\
\mu_i &= w_i - (w_i)^{-1},
\end{aligned}
$$

then we have

Proposition 1.2: *Let* π *be a finite* 2-*group, then*

$$Q(\zeta_3)\pi = \Sigma M_{n_i}(\mathbb{F}_i)$$

where $\mathbf{F}_i$ *is one of the three fields* $Q(\zeta_3,\lambda_i)$, $Q(\zeta_3,\mu_1)$, $Q(\zeta_3,w_i)$.

We must be careful about the involution when we use the decomposition of 1.2. Every involution τ on $M_{n_i}(\mathbb{F}_i)$ agreeing with complex conjugation on the center $\mathbf{F}_i$ is equivalent to the usual involution — conjugate transpose,

$$M \to \overline{M}, \quad m_{ij} \to \overline{m}_{ji}.$$

Precisely, there is a matrix S so that

$$S\overline{M}S^{-1} = \tau(M).$$

Moreover, $\tau(S) = \pm S$, and we say that τ has type IIa if S can be chosen so that $\tau(S) = S$. Otherwise, τ has type IIb.

Lemma 1.3: $L_i(M_{n_i}(\mathbf{F}_i), \tau) = L_{i+2}(M_{n_i}(\mathbb{F}_i), {}^-)$ *if* τ *has type* IIb. *Otherwise* $L_i(M_{n_i}(\mathbf{F}_i), \tau) = L_i(M_{n_i}(\mathbf{F}_i), {}^-)$.
(See eg. [2] for a discussion.)

In the current case each simple summand of $Q\pi$ is invariant under the involution so the same is true for each summand of $Q(\zeta_3)\pi$. Moreover, each summand of $Q\pi$ of the form $M_{n_i}(Q(\lambda_i))$, $M_{n_i}(Q(\mu_i))$, or $M_{n_i}(Q(\zeta_{2^i}))$, on tensoring with $Q(\zeta_3)$ gives rise to a IIa summand in $Q(\zeta_3)\pi$ while each of the summands of type $M_{n_i}(D_i)$ gives rise to a type IIb summand in $Q(\zeta_3)\pi$. Thus, the distinction between the quaternion and matrix representations in $Q(\zeta_3)\pi$ appears only in the shifting of their contributions to $L_*(Q(\zeta_3)\pi)$ according to 1.3.

It should also be noted that 1.3 holds at all the completions of $\mathbb{F}_i$, and that S above will be integral over $Z(1/6,\zeta_3)\pi$. Consequently, the Meyer-Vietoris sequence below makes the L_*^p calculations routine.

1.4
$$\cdots \longrightarrow L^h_{k+1}(\hat{Q}_2(\zeta_3)\pi) \oplus L^p_{k+1}(\hat{Q}_3(\zeta_3)\pi) \longrightarrow L^p_k(Z(\zeta_3)\pi) \longrightarrow$$
$$L^h_k(\hat{Z}_2(\zeta_3)\pi) \oplus L^p_k(\hat{Z}_3(\zeta_3)\pi) \oplus L^h_k(Z(1/6,\zeta_3)\pi) \longrightarrow L^h_k(\hat{Q}_2(\zeta_3)\pi) \oplus L^p_k(\hat{Q}_3(\zeta_3)\pi) \longrightarrow \cdots$$

The next step is to extend the calculations above to the $L^h_*(Z(\zeta_3)\pi)$ groups. That we can do this is ultimately the reason we are able to prove the results of the introduction, and it rests on the fact that the involution is non-trivial on the center of $Z(\zeta_3)\pi$.

Lemma 1.5: *The involution on* $\tilde{K}_0(Z(\zeta_3)\pi)$, $[P] \rightleftarrows [P^*]$, *is the image of the involution on* $K_1(\hat{Q}_2(\zeta_3)\pi)$ *given by*

$$\{M_{i,j}\} \rightleftarrows \{(M_{j,i}{}^*)^{-1}\},$$

where $M_{i,j}$ *is a non-singular matrix with coefficients in* $\hat{Q}_2(\zeta_3)\pi$.

Definition 1.6: *The image of* $K_1(\hat{Z}_2(\zeta_3)\pi)$ *in* $K_1(\hat{Q}_2(\zeta_3)\pi)$ *is denoted* $K_1'(\hat{Z}_2(\zeta_3)\pi)$. *The quotient* $UK_1(\hat{Q}_2(\zeta_3)\pi)/K_1'(\hat{Z}_2(\zeta_3)\pi)$ *is written* $K_0'(Z(\zeta_3)\pi)$ *where* $UK_1(\hat{Q}_2(\zeta_3)\pi)$ *is the set of units in* $K_1(\hat{Q}_2(\zeta_3)\pi) = \Sigma\hat{Q}_2(\zeta_3,v_i)$, *that is*

$$UK_1(\hat{Q}_2(\zeta_3)\pi) = \Sigma U\hat{Q}_2(\zeta_3,v_i).$$

We have (extending an argument of Oliver [11])

Theorem 1.7: *Let* π *be a finite* 2-*group, then*

$$\hat{H}_k(Z/2;\ K_0'(Z(\zeta_3)\pi)) = 0$$

for all k.

We introduce the notation

$\lambda(\pi)$ = number of $M_{n_i}(Q(\lambda_i))$ summands of $Q\pi$,

1.8 $d(\pi)$ = number of $M_{n_i}(D_i)$ summands of $Q\pi$,

$c(\pi)$ = number of conjugacy classes in π,

and taking account of 1.2, 1.3, 1.7 we obtain

Theorem 1.9: *The groups* $L^h_*(Z(\zeta_3)\pi)$ *are given as follows:*

$$L^h_0(Z(\zeta_3)\pi) \cong Z^{c(\pi)} \oplus (Z/2)^{d(\pi)} \oplus V_0.$$
$$L^h_1(Z(\zeta_3)\pi) \cong (Z/2)^{rk(\pi^{ab})} \oplus V_1.$$
$$L^h_2(Z(\zeta_3)\pi) \cong Z^{c(\pi)} \oplus (Z/2)^{\lambda(\pi)-rk(\pi^{ab})-d(\pi)-1} \oplus V_2.$$
$$L^h_3(Z(\zeta_3)\pi) \cong (Z/2)^{d(\pi)}.$$

Here, the groups $V_0 = V_2 = (Z/2)^{n(\pi)-\alpha(\pi)}$ are not in the image of $\Omega_*(B_\pi)$. Also, the subgroup

$$(Z/2)^{rk(\pi^{ab})} \quad \text{in} \quad L_1()$$

comes from $L_1^p()$ and (except for a $Z/2$ kernel) project non-trivially to corresponding elements for $L_1^p(Z\pi^{ab})$, while the elements in $L_3^h()$ come from the $Z/2$'s in $\hat{H}_*(Z/2; \tilde{K}_0(Z(\zeta_3)\pi))$ associated to the modules W_i at the quaternion representations in $Q(\zeta_3)\pi$. Finally, the elements in V_1 go to 0 under τ_K.

Remark 1.10: Theorem B is essentially immediate from 1.9. Moreover, it is almost as easy to obtain Theorem C(b), (d) from 1.9. To obtain the remaining results and the explicit characteristic class formulae we must analyze the elements above with more care. This requires the introduction of L-groups with coefficients.

Section 2. L *theory with coefficients.*

A $Z/2$ manifold consists of a pair $\{M, f: M \longrightarrow S^1\}$, where f is defined up to homotopy and $f^*(\iota)$ is an integral lifting of the first Stiefel-Whitney class of M, w_1. If $\{M^n, f\}$ and $\{N^s, g\}$ are $Z/2$ manifolds then the product $\{M^n \times N^s, fp_1 + gp_2\}$ (p_i projection onto the ith factor) is again a $Z/2$ manifold.

The $Z/2$ bordism group of the space X, denoted $\Omega_*(X, Z/2)$, is the set of bordism classes of maps $F: M^n \longrightarrow X$ where M^n is a closed $Z/2$ manifold. Product induces an associative pairing

$$\mu_{X,Y}: \Omega_*(X, Z/2) \otimes \Omega_*(Y, Z/2) \longrightarrow \Omega_*(X \times Y, Z/2)$$

so $\Omega_*(pt, Z/2) = \mathbf{N}_*$ becomes a graded ring with unit and $\Omega_*(X, Z/2)$ is a graded module over $\mathbf{N}_*$. There is a natural augmentation $\varepsilon: \mathbf{N}_* \longrightarrow Z/2$ with kernel the ideal $\mathbf{I} = \bigoplus_{j>0} \mathbf{N}_j$. There is a Hurewicz homomorphism

2.1 $$\hat{h}: \Omega_*(X, Z/2) \otimes_{\mathbf{N}_*} ((\mathbf{N}_*/\mathbf{I}) = Z/2) \longrightarrow H_*(X, Z/2)$$

and $\hat{h}$ is an isomorphism. In particular,

$$\Omega_*(X, Z/2) \cong \mathbf{N}_* \otimes_{Z/2} H_*(X, Z/2)$$

$$\Omega_*(X \times Y, Z/2) \cong \Omega_*(X, Z/2) \otimes_{\mathbf{N}_*} \Omega_*(Y, Z/2)$$

There is also an exact sequence

2.3 $$\cdots \longrightarrow \Omega_n(X) \xrightarrow{\times 2} \Omega_n(X) \longrightarrow \Omega_n(X; Z/2) \xrightarrow{\partial} \Omega_{n-1}(X) \xrightarrow{\times 2} \cdots$$

where $\Omega_n(X)$ is the usual oriented bordism group of X.

The most important group for our applications is $Z/2$ itself. $B_{Z/2}$ is an abelian H-space and the multiplication makes both its $Z/2$ bordism and homology into graded rings. Indeed,

2.2 $$H_*(B_{Z/2};\ Z/2) \cong E(e_1)\otimes E(e_2)\otimes E(e_4)\otimes \ldots E(e_{2^i}) \ldots$$

so $\Omega_*(B_{Z/2};\ Z/2) = \mathbf{N}_*\otimes E(e_1, \ldots, e_{2^i} \ldots)$ as a graded ring.

We need explicit generators for these bordism rings. $Z/2$ acts on RP^n by

$$T(x_0, \ldots, x_n) = (x_0, \ldots, x_{n-1}, -x_n).$$

T is orientation reversing when n is odd. Set

2.5 $$M^{2n} = S^1\times RP^{2n-1}/\{(y, x) = (-y, Tx)\}.$$

The first of these, $M^2 = S^1\times_T S^1$ is just the Klein bottle, with fundamental group $Z\times_T Z = \{T,\tau|\tau T\tau^{-1} = T^{-1}\}$. The remaining manifolds all have abelian fundamental group $(Z/2)\times Z$.

There are maps $f_{2n}\colon M^{2n} \longrightarrow B_{Z/2}$ taking the first generator T to the generator of $Z/2$ and the second generator to 1. These pairs $\{M^{2n}, f_{2n}\}$ generate $\Omega_*(B_{Z/2};\ Z/2)$ as a module over $\mathbf{N}_*$, and, from 2.4, we have

Proposition 2.6: *Let* $M^1 = S^1$, *and* $M(i) = M^{2^i}$. *Then for each of the spaces* $M(i)$, $i = 0, 1, 2, \ldots$ V^2, $VSq^1(V)$, $(Sq^1V)^2$ *are all just the class* 1 *in dimension* 0, *and these* $M(i)$ *generate* $\Omega_*(B_{Z/2};\ Z/2)$ *as a ring over* $\mathbf{N}_*$.

There is a natural orientation

2.7 $$\hat{\mu}\colon\ \Omega_*(B_\pi;\ Z/2) \xrightarrow{\hat{\mu}} L^h_*(Z(\zeta_3)\pi;\ Z/2) \longrightarrow L^h_*(Z(\zeta_3)\pi\times Z^-).$$

For A any ring with involution τ we have the exact sequence [9]

2.8 $$\ldots \longrightarrow L^p_*(A,\tau) \xrightarrow{2} L^h_*(A,\tau) \longrightarrow L^h_*(A[t,t^{-1}]^-) \longrightarrow$$
$$L^p_{*-1}(A,\tau) \xrightarrow{2} L^h_{*-1}(A,\tau) \longrightarrow \ldots$$

2.8 is extremely efficient for calculating the $L^h_*(Z(\zeta_3)\pi\times Z^-)$.

The L groups $L^h_n(A;\ Z/2)$ are defined in terms of models which consist of n-dimensional quadratic Poincaré complexes D with $\partial D = \delta D\oplus\delta D$ where δD is a closed Poincaré complex of dimension $n-1$. Bordism is defined in the evident way and there is an exact sequence

2.9
$$\cdots \longrightarrow L_n^h(A) \xrightarrow{\times 2} L_n^h(A) \longrightarrow L_n^h(A;\ Z/2) \longrightarrow L_{n-1}^h(A) \xrightarrow{\times 2} L_{n-1}^h(A) \longrightarrow \cdots$$

2.8, 2.9 fit together to give the commutative diagram

2.10
$$\begin{array}{ccccccc} \cdots \longrightarrow & L_*^h(A) & \xrightarrow{\times 2} & L_*^h(A) & \longrightarrow & L_*^h(A;\ Z/2) & \longrightarrow \cdots \\ & \downarrow & & \downarrow & & \downarrow v & \\ \cdots \longrightarrow & L_*^p(A) & \xrightarrow{\times 2} & L_*^h(A) & \longrightarrow & L_*^h(A[t,t^{-1}]^-) & \longrightarrow \cdots \end{array}$$

We use the map v in 2.10 when $\tilde{K}_0(Z(\zeta_3)\pi) = 0$ (e.g. when $\pi = Z/2$ or $Z/4$) to study the map

$$\hat{\mu}:\ \Omega_*(B_\pi;\ Z/2) \longrightarrow L^h(Z(\zeta_3)\pi;\ Z/2).$$

There is a product pairing

2.11
$$L_h^i(Z;\ Z/2) \otimes L_j^h(Z(\zeta_3)\pi;\ Z/2) \xrightarrow{\cup} L_{i+j}^h(Z(\zeta_3)\pi;\ Z/2)$$

and from ([12], page 173) we have

$$L_h^j(Z;\ Z/2) = L_h^j(Z(Z^-)) = \begin{cases} Z/2 & \text{generator } I,\ j \equiv 0 \bmod(4) \\ Z/2 & \text{generator } \gamma,\ j \equiv 1 \bmod(4) \\ Z/2 & \text{generator } \tau,\ j \equiv 2 \bmod(4) \\ 0 & j \equiv 3 \bmod(4) \end{cases}.$$

Moreover, from [4] or [10], the map

$$\mu:\ \mathbf{N}_* \longrightarrow L_h^*(Z)$$

is given by the formula

$$\mu\{M\} = \langle V^2, [M]\rangle I + \langle VSq^1V, [M]\rangle\gamma + \langle (Sq^1V)^2, [M]\rangle\tau.$$

Thus, the general form of the evaluation formulae similar to that given in [13], is

Corollary 2.12: *There are well-defined elements* $\iota_{*,\kappa}, \gamma_{*,\kappa}, \tau_{*,\kappa} \in H^*(B_\pi;\ Z/2)$ *which have the property that the image of* $\{M, f\} \in \Omega_*(B_\pi, Z/2)$ *in* $L_*^h(Z(\zeta_3)\pi;\ Z/2)$ *is given by a collection of formulae*

$$(\langle V^2f^*(\iota_{*,\kappa}),[M]\rangle + \langle VSq^1Vf^*(\gamma_{*,\kappa}),[M]\rangle + \langle (Sq^1V)^2f^*(\tau_{*,\kappa}),[M]\rangle)\kappa,$$

one for each generator κ *in* $L_*^h(Z(\zeta_3)\pi,\ Z/2)$.

In our case we have

Theorem 2.13: *The maps*

$$\mu(\gamma^*\ ): \Omega_*(B_\pi) \longrightarrow L^h_{*+1}(Z(\zeta_3)\pi)$$
$$\hat{\mu}(\gamma^*\ ): \Omega_*(B_\pi,\ Z/2) \longrightarrow L^h_{*+1}(Z(\zeta_3)\pi;\ Z/2)$$

are identically zero while the map

$$\mu(\tau^*\ ): \Omega_*(B_\pi,\ Z/2) \longrightarrow L^h_{*+2}(Z(\zeta_3)\pi;\ Z/2)$$

is multiplication by the non-zero element in $L^h_2(Z(\zeta_3);\ Z/2)$.

Remark 2.14: 2.13 implies that the only coefficients appearing in the general product formulae are $<(V)^2 f^*(\kappa),\ [M]>$ or $<(Sq^1 V)^2 f^*(\kappa'),\ [M]>$.

Section 3. *The products* $S^1\times_T RP^{4n-1}$

An equivariant cell decomposition of the universal cover $R\times S^{4n-1}$ of $S^1\times_T RP^{4n-1}$ is given as the product of the usual decomposition of S^{4n-1} into + and - hemispheres and the line into the union of its integer points and the intervals $[m,m+1]$, as m runs over the integers. On cells of the form $e_1\otimes e_j$ the operation of the fundamental group is just the product action as long as $j\neq 4n-1$, but for $j = 4n-1$, $T(e_i\times(e_{4n-1}{}^+)) = (e_i)\times e_{4n-1}{}^-$, while $\tau(e_i\otimes(e_{4n-1}{}^+)) = -\tau(e_i)\times(e_{4n-1}{}^-)$.

In this section we study the class of the image $<Z,-(\zeta_3)>\otimes\{\mathbf{C},\ \phi\}$ in $L^h_0(Z(\zeta_3)Z/2;\ Z/2)$, where $\{\mathbf{C},\ \phi\}$ is the Ranicki symmetric structure induced on the complex $S^1\times_T RP^{4n-1}$ via a suitable diagonal approximation. An equivariant diagonal approximation on $\mathbf{C}(S^{4n-1})$ is given by

3.1 $$\tilde{\Delta}_\#(e_i{}^+) = \Sigma e_j\otimes(T^j)e_{i-j}.$$

and a Z equivariant diagonal approximation on R is given by

$$e_0 \longrightarrow e_0\otimes e_0,\quad e_1 \longrightarrow e_0\otimes e_1 + e_1\otimes\tau e_0.$$

On tensoring these maps together we get a suitable diagonal approximation on the $4n-2$ skeleton. On the $4n-1$ skeleton, the same formula works as we easily check, and on the $4n$ skeleton a long but direct calculation gives

$$\Delta(\partial e_1\times e_{4n-1}{}^+) = \partial\{1\times T\tau(e_1\times e_{4n-1}{}^+\otimes 1) + 1\otimes e_1\times e_{4n-1}{}^+ +$$
$$\sum_{n-1>i>1} e_1\times e_i\otimes T^i\tau e_0\times e_{4n-i-1} + \Sigma(-1)^i e_0\times e_i\otimes T^i e_1\times e_{4n-i-1} -$$
$$\sum_{0<i<n} \tau e_0\times e_i\otimes\tau T^i e_0\times e_{4n-i-1} - e_1\times e_0\otimes T\tau e_0\times e_{4n-1}{}^+ -$$
$$e_0\times e_{4n-1}{}^+\otimes T e_1\times e_0 + \tau T\times\tau T(e_0\times e_{4n-1}{}^+\otimes e_0\times e_1)\}.$$

Using this and calculations modeled on those of [7], we have

Theorem 3.2: *The class of* $\{\mathbf{C}^{4n-*} \longrightarrow \mathbf{C}_*\} \otimes \langle 1, \zeta_3\rangle$ in $L^h_{4n}(Z(\zeta_3)Z^- \times Z/2)$ *is the image of the class* $\langle 1\rangle \perp \langle T\rangle$ *coming from* $L^h_{4n}(Z(\zeta_3)Z/2)$ *under the usual inclusion.*

The case of $M(1)$ is slightly more delicate. $Z\times_T Z = \{T,\tau \mid \tau^{-1}T\tau T = 1\}$, and a resolution is given by the complex

3.3 $$C[A] \longrightarrow C[e] \oplus C[f] \longrightarrow C[1]$$

where

3.4 $$\begin{aligned} \partial[e] &= (T-1)[1] \\ \partial[f] &= (\tau-1)[1] \\ \partial[A] &= (T^{-1}+\tau^{-1})[e] + (\tau^{-1}T-\tau^{-1})[f]. \end{aligned}$$

We define an involution on the group ring $Z(Z\times_T Z)$ by $T \rightleftarrows T^{-1}$, $\tau \rightleftarrows (-\tau^{-1})$. Then we proceed as before constructing an explicit quadratic form associated to this group, resolution, and involution. The result is the skew symmetric quadratic form with matrix

3.5 $$Q_{Klein} = \begin{pmatrix} \zeta^2(T^{-1}-1)-\zeta(T-1) & T+\zeta^2(1+T^2\tau^{-1})(T-1) \\ -T^{-1}-\zeta(1-T^{-2}\tau)(T^{-1}-1) & -(\zeta T^{-2}\tau+\zeta^2T^2\tau^{-1}) \end{pmatrix}$$

Now, consider the surjection $\nu\colon Z\times_T Z \longrightarrow Z/2$ defined by $\nu(T) = T$, $\nu(\tau) = 1$. This gives an involution preserving homomorphism $\hat{\nu}\colon Z\times_T Z \longrightarrow Z/2\times Z^-$, and we have, after modifying the form slightly

$$\hat{\nu}(Q_{Klein}) = \begin{pmatrix} (\zeta^2-\zeta)(T-1) & T+(\zeta^2+2\tau^{-1})(T-1) \\ -T-(\zeta+2\tau)(T-1) & -(\tau+\tau^{-1}+\zeta^2-\zeta)(T-1) \end{pmatrix}.$$

In particular this implies

Proposition 3.6: $\langle T\rangle \otimes \hat{\nu}(Q_{Klein}) \sim \langle -1\rangle \otimes (\hat{\nu}(Q_{Klein})$.

Finally, using the product operation in $\Omega_*(B_{Z/2}; Z/2)$, we have

Corollary 3.7: *The product map*

$$\gamma\colon \{S^1\times_T S^1\} \times \{S^1 \times_T S^1\} \longrightarrow B_{Z/2}\times B_{Z/2} = B_{Z/2\times Z/2}$$

is represented in $L^h_0(Z(\zeta_3)Z/2\times Z/2;\ Z/2)$ *either by* 0 *or by the image of the class* $\langle 1\rangle \mid \langle T_1\rangle \perp \langle T_2\rangle \perp \langle T_1T_2\rangle$ *in* $L^h_0(Z(\zeta_3)Z/2\times Z/2)$.

Section 4. *The proof of the theorem* C.

The following result allows us to concentrate on multiplicative generators.

Lemma 4.1: *Let* $(\mathbf{C},\phi_*)$, $(\mathbf{D},\phi_*')$ *be two symmetric Poincaré Duality complexes of even dimensions* 2n, 2m *respectively. Then*

$$\mu(\mathbf{C}\otimes\mathbf{D},\phi_*\otimes\phi_*') = -g(\mu(\mathbf{C},\phi_*)\otimes\mu(\mathbf{D},\phi_*'))$$

where $g\colon Z(\zeta_3)\longrightarrow Z(\zeta_3)$ *is the Galois automorphism.*

For $\pi = Z/2$ we need only consider the images $\mu(S^1)$, $\mu(S^1\times_T S^1)$, $\mu(S^1\times_T RP^3)$, ... $\mu(S^1\times_T RP^{2^i-1})$ ..., and we have

Lemma 4.2: *Each of the generators above has non-trivial image in* $L_0^h(Z(\zeta_3)Z/2;\ Z/2)$, *but every product maps to* 0.

Corollary 4.3: *The* κ's *for* $B_{Z/2}$ *are non-zero only in dimension* 2^i, *for* $i = 0,1,2,\ldots$.
(Since the model manifolds have trivial Wu-product classes V^2, $\Sigma V_i Sq^1(V_i)$, $(Sq^1V)^2$ it follows that the homomorphism into $L_2^h(Z(\zeta_3)Z/2;\ Z/2)$ directly calculates the κ's. See e.g. [13], [4], [14], and especially Theorem 2.13.)

Our result is not quite so complete for the group $Z/2\times Z/2$.

Lemma 4.4: *For the group* $Z/2\times Z/2$ *we have*

a) $\mu(M(i)\times 1)$, $\mu(1\times M(1))$, $\mu(M(0)\times M(0))$ *are all non-zero.*

b) $\mu(M(i)\times M(j)) = \mu(M(j)\times M(i))$ *all* i, j.

c) $\mu(M(0)\times M(i)) = 0$ *for* $i>1$, *while*
$\mu(M(i)\times M(j)) = \langle 1\rangle\perp\langle T_1\rangle\perp\langle T_2\rangle\perp\langle T_1T_2\rangle = V \neq 0$ *for* $i,j > 1$.

d) $\mu(M(1)\times M(1))$ *is either* 0 *or* V.

Note that this does not completely determine the image of the obstructions but it does determine them up to classes which are in the image of the Steenrod algebra action on the classes in dimensions one and two.

For the dihedral groups $D(2^i,2)$ we have

Lemma 4.5: *There are two inclusions* $\phi_j^i\colon Z/2\times Z/2\longrightarrow D(2^i,2)$ *which surject in* mod(2) *homology and for the* mod(2) L-*groups satisfy* $(\phi_j^i)(V)\neq 0$ *in* $L_0^h(Z(\zeta_3)D(2^i,2);\ Z/2)$. (V is defined in 4.4(c).)

We must also consider the realization question for the dihedral groups. The generators for the subgroup

$$Z/2^{\lambda(\pi)-rk(\pi^{ab})-d(\pi)-1}$$

in $L_2^h(Z(\zeta_3)\pi)$ either project non-trivially to π^{ab} or are induced up from dihedral subquotients. In turn, on dihedral groups, since the mod(2) homology is entirely in the image from two copies of $Z/2\times Z/2$, it follows that realization is determined by analyzing the group $Z/2\times Z/2$.

Proposition 4.6: $L_2^h(Z(\zeta_3)D(2^i,2)) = (Z/2)^2 \oplus Z^5$, *with one of the* $Z/2$'s *surjecting onto the* $Z/2$ *in* $L_2^h(Z(\zeta_3)Z/2\times Z/2)$.

It is easy to show that when $k=2$ the induction maps $\phi_j^{i,!}$ are both non-trivial on the torsion class, mapping it onto the kernel of projection onto $D(2^i,2)^{ab}$. Thus it follows that each of the induced up $Z/2$'s in $L_2^h(Z(\zeta_3)\pi)$ is realized at least at some point by a non-trivial example. Of course, this is not to assert that the element is always non-trivial, but if it lifts non-trivially to the subgroup (of which the dihedral group is the quotient) then it will induce non-trivially to π.

The classes in the torsion subgroup of $L_0^h(Z(\zeta_3)\pi)$ are all induced from quaternion subquotients. Consequently, in order to show that the image of $\Omega(B_\pi)$ in $L_0^h(Z(\zeta_3)\pi)$ is just the Z generated by $\langle 1\rangle$ it suffices to show that the same is true for the quaternion groups $Q(2^i)$. But for the quaternion group we can make an explicit calculation as follows

Lemma 4.7: *There exists a set of generators for the* mod(2) *bordism of* $Q(2^i)$ *as a module over* $\mathbf{N}_*$ *having trivial* V^2, VSq^1V, *and* $(Sq^1V)^2$ *genus, which all have trivial images in* $L_*^h(Z(\zeta_3)Q(2^i);\ Z/2)$ *if their dimension are* 4 *or greater, while the generators of dimension* < 4 *all have non-trivial (and independent) images.*

To complete the proof we need to consider the extra terms $\langle (Sq^1V)^2 f^*(\kappa'), [M]\rangle$ which can appear in the product formulae for the map μ as coefficients. For oriented M it suffices to analyze the situation for the model groups, and we have

Theorem 4.8: *For the model groups the coefficients above are identically* 0 *whenever* M *is an oriented manifold.*

Proof: For $\pi = Z/2$ or $Z/4$ we know that $L_3^h(Z(\zeta_3)\pi;\ Z/2) = 0$, so $\mu(\tau\cdot M(i)) = 0$ in both these cases. Likewise, when $i > 1$, 4.1 implies $\mu(\tau\cdot M(i))$ will involve detection by coefficients $\langle Sq^1V)^2 f^*(b), [M]\rangle$, and since $Sq^1(b) = 0$, it follows that

4.9 $$(Sq^1V)^2f^*(b) = Sq^1(V\cdot Sq^1V\cdot f^*(b))$$

and so will vanish in top dimension on any oriented manifold.

When $\pi = Z/2\times Z/2$ we have that $\mu(\tau\cdot M(1)\times M(1))$ is represented by either $\tau\cdot V$ or 0. But 4.1 implies that $\tau\cdot V = 0$ so this class vanishes. Otherwise,

$$\mu(\tau\cdot(M(i)\times M(j))) = \mu((\tau\cdot M(i))\times M(j)) = \mu(M(i)\times(\tau\cdot M(j))$$

and unless both $i, j = 1$ the proof for $Z/2$ implies that the above class gives 0 in $L_*^h(Z(\zeta_3)Z/2\times Z/2;\ Z/2)$.

When $\pi = Q(2^i)$ we dispose of the classes in dimension 1 by noting that they factor through $Z/4$, hence $\mu(\tau\cdot M(0))$ represents 0 in $L_3^h(Z(\zeta_3)Q(2^i);\ Z/2)$. The remaining classes in dimensions 2 and 3 on producting with τ give formulae involving only the classes in dimensions 2 and 3 in $H^*(Q(2^i);\ Z/2)$, but these classes are all in the kernel of Sq^1, so 4.9 implies the vanishing of these classes on oriented manifolds and 4.8 follows.

BIBLIOGRAPHY

1. F. Clauwens, "The K-theory of almost symmetric forms", *Topological structures II*. 1979. Mathematical Centre Tracts, Proc. of the symposium in Amsterdam. October 31 - November 2, 1978 115(1979), 41-49.

2. I. Hambleton, R. J. Milgram, "The surgery obstruction groups for finite 2-groups", Invent. Math., 63(1980), 33-52.

3. I. Hambleton, R. J. Milgram, L. Taylor and B. Williams, "Surgery with finite fundamental group, III: Twisted Kervaire Problem", (to appear).

4. R. J. Milgram, "Surgery with coefficients", Ann. of Math., 2(100) (1974), 194-248.

5. R. J. Milgram, "Surgery with finite fundamental group I: the obstructions", (mimeo. U.C.S.D., May 1985).

6. R. J. Milgram, "Surgery with finite fundamental group II: the oozing conjecture", (mimeo. U.C.S.D., May 1985).

7. R. J. Milgram, "The Cappell-Shaneson example", Proc. of L-Theory Conf.: Rutgers, 1983, Springer-Verlag Lecture Notes in Mathematics, (to appear).

8. R. J. Milgram, A. Ranicki, "Some product formulae for nonsimply connected surgery problems", mimeo. (1985).

9. R. J. Milgram, A. Ranicki, A. Wadsworth, "Witt rings and reciprocity laws for Genus 0 function fields", (to appear).

10. J. Morgan, D. Sullivan, "The transversality characteristic class and linking cycles in surgery theory", Ann. of Math., 99(1974), 463-544.

11. R. Oliver, "SK_1 for finite group rings II", Math. Scand. 47(1980), 195-231.

12. A. Ranicki, "The algebraic theory of surgery I, II", Proc. Lond. Math. Soc. 3(40)(1980), 87-283.

13. L. Taylor, B. Williams, "Surgery spaces, formulae and structure", Proceedings 1978 Waterloo Conference on Algebraic Topology, Springer-Verlag Lecture Notes in Mathematics #741, 170-195.

14. C.T.C. Wall, "Formulae for surgery obstructions", Topology 15(1976), 189-210.

Contemporary Mathematics
Volume **58**, Part II, 1987

KO(B)-GRADED STABLE COHOMOTOPY OVER B AND RO(G)-GRADED G-EQUIVARIANT STABLE COHOMOTOPY: A FIXED POINT THEORETICAL APPROACH TO THE SEGAL CONJECTURE

CARLOS PRIETO

ABSTRACT. The concept of a KO(B)-graded cohomology theory for spaces over B is introduced in analogy to RO(G)-graded G-equivariant cohomology theories. Using methods of fixed-point-theory and coincidence-theory, stable cohomotopy over B as well as G-equivariant stable cohomotopy are studied. We describe a delate natural transformation from the suitably completed RO(G)-graded G-equivariant stable cohomotopy to the KO(BG)-graded stable cohomotopy over B and show it is an isomorphism when G is a finite group.

§0. INTRODUCTION

In [Pr] we defined the groups $FIX^k(X, A)$ of "k-fixed point situations"

0.1
$$\begin{array}{ccc} E \supset V & \xrightarrow{f} & \mathbb{R}^k \times E \\ & \searrow \quad X \quad \swarrow & \end{array}$$

if $k \geqslant 0$, resp.

0.2
$$\begin{array}{ccc} \mathbb{R}^{-k} \times E \supset V & \xrightarrow{f} & E \\ & \searrow \quad X \quad \swarrow & \end{array}$$

if $k < 0$, where $E \to X$ is an ENR_X (see [Do] §1) and such that the k-fixed point set of f

$$Fix^k(f) = \{v \in V | f(v) = (0, v)\}, \quad \text{resp.} \quad = \{v = (y, e) \in V | f(y, e) = e\}$$

lies properly over X and its image in X does not intersect A.

Let ϵ^k denote the trivial vector bundle (over X), then

$$\epsilon^k \times_X E = \mathbb{R}^k \times E.$$

This remark suggests that we consider maps of the following type:

0.3
$$\begin{array}{ccc} E \supset V & \xrightarrow{f} & \xi \underset{X}{\times} E \\ & \searrow \quad X \quad \swarrow & \end{array}$$

resp.

0.4
$$\begin{array}{ccc} \eta \underset{X}{\times} E \supset V & \xrightarrow{f} & E \\ & \searrow \quad X \quad \swarrow & \end{array}$$

or better

0.5
$$\begin{array}{ccc} \eta \underset{X}{\times} E \supset V & \xrightarrow{f} & \xi \underset{X}{\times} E \\ & \searrow \quad X \quad \swarrow & \end{array}$$

where ξ and η are (real) vector bundles over X, and $E \to X$ is again an ENR_X. These maps are to be *compactly* (ξ,η)-*fixed over* (X, A), i.e., that the (ξ,η)-*fixed point set* of f,

$$\mathrm{Fix}^{\xi,\eta}(f) = \{v = (y, e) \in V | f(y, e) = (0, e)\}$$

lies properly over X and its image in X does not meet A.

As in [Pr] a suitable equivalence relation between two such maps is introduced (see §2) and the set of equivalence classes

$$\mathrm{FIX}^{\xi,\eta}(X, A)$$

has the structure of an abelian group, the sum given by disjoint union.

If for example, X is a space over B (i.e., $p: X \to B$) and $\xi,\eta \downarrow B$ are vector bundles, we can repeat this construction but now for the induced bundles $p^*\xi$, $p^*\eta \downarrow X$, and still denote the corresponding group by $\mathrm{FIX}^{\xi,\eta}(X, A)$. If $[\xi]-[\eta] = [\xi']-[\eta'] \in KO(B)$, then one has

$$\mathrm{FIX}^{\xi,\eta}(X, A) \cong \mathrm{FIX}^{\xi',\eta'}(X,A)$$

(see TH 2.6), so that we have a well defined group

$$\mathrm{FIX}_B^b(X, A)$$

for each $b \in KO(B)$. These functors FIX_B^b have the structure of a cohomology theory for spaces over B which is graded by the elements of $KO(B)$ in the same way that G-equivariant cohomology theories which are graded by the real representation ring $RO(G)$ of G.

In fact, the fixed point theory methods allow us to define in general, for G a compact lie group and B a G-space, a cohomology theory $G\text{-}\mathrm{FIX}_B^*$ for G-spaces over B graded by $KO_G(B)$, the equivariant real K-theory of B.

In [Do] and [Pr] it is proved that the groups $\mathrm{FIX}^k(X, A)$ (for $k=0$ and $A=\phi$ in [Do]) are naturally isomorphic to the stable cohomotopy groups

$$\pi_S^k(X, A) := \operatorname{colim} [\Sigma^j(X \cup CA), \mathbb{S}^{k+j}],$$

where $X \cup CA$ denotes the cone of the inclusion $A \subset X$.

Thus, the groups $\mathrm{FIX}_B^b(X, A)$ determine a generalization of stable cohomotopy for pairs of spaces (X, A) over B which in turn can be directly defined by the formula 1.11. See 2.24.

In this paper we shall introduce the definition of a $KO(B)$-graded cohomology theory over B and the more general KO-graded parametrized cohomology (relating several base-spaces B). In the second paragraph we shall discuss FIX_B^*. In the third paragraph the definition of G-FIX^* is given and it is mentioned how to prove that this theory is isomorphic to equivariant stable cohomotopy as done also by Ulrich in his thesis [Ul] (see 3.18). In the same paragraph we construct, given a principal G-bundle $P \to B$ a natural transformation

$$G\text{-}\mathrm{FIX}^*(X, A) \to \mathrm{FIX}_B^*(P \times_G X, \ P \times_G A),$$

which for $EG \to BG$ extends Segal's homomorphism $A(G) \to \pi_S^0(BG)$, (modulo the corresponding isomorphism).

We finish by proving an extension of Segal's conjecture which states that the appropriately completed G-FIX -theory is isomorphic to the FIX_{BG}^* -theory.

I want to thank A. Dold and B. Schäfer for having suggested the study of coincidence situations 0.3 - 0.5.

§1. K-GRADED PARAMETRIZED COHOMOLOGY

For the sake of clarity I shall first define the concept of $KO(B)$-graded cohomology over B.

1.1 Let $\mathfrak{W}_B^2$ be a category of pairs of topological spaces over B, $A \subset X \to B$. A $KO(B)$-*graded cohomology theory over* B, or briefly, a *cohomology over* B consists of a family of contravariant functors

$$h_B^b : \mathfrak{W}_B^2 \to \mathfrak{Ab}$$

($\mathfrak{Ab}$ = an abelian category) and a family of natural transformations

$$\delta = \delta^b : h_B^b \circ T \to h_B^{b+1}$$

for all $b \in KO(B)$, where $T: \mathfrak{W}_B^2 \to \mathfrak{W}_B^2$ is the functor sending (X, A) to $(A, \phi) = A$. These families satisfy the following axioms.

1.2 HOMOTOPY: Let $A \subset X \to B$ and $A' \subset X' \to B$ be objects in $\mathfrak{W}_B^2$ and let f_0, f_1: $(X', A') \to (X, A)$ be fiberwise homotopic maps, then

$$f_0^* = f_1^* : h_B^b(X, A) \to h_B^b(X', A')$$

for all $b \in KO(B)$.

1.3 EXCISION: Let $A \subset X \to B$ be an object in $\mathfrak{W}_B^2$ and let C be such that $\overline{C} \subset \mathring{A}$ and $A-C \subset X-C \to B$ is also an object in $\mathfrak{W}_B^2$, then the canonical homomorphism

$$h_B^b(X, A) \to h_B^b(X-C, A-C)$$

is an isomorphism for all $b \in KO(B)$.

1.4 EXACTNESS: For every $b \in KO(B)$ and every pair (X, A) in $\mathfrak{W}_B^2$, the sequence

$$\ldots \to h_B^b(X) \to h_B^b(A) \overset{\delta}{\to} h^{b+1}(X, A) \to h^{b+1}(X) \to \ldots$$

is exact.

1.5 These axioms, which correspond to the usual axioms of a $\mathbb{Z}$-graded cohomology theory over B, tell us that a cohomology over B is just a family of "usual" cohomologies over B, one for each $b \in KO(B)$. The next axiom, suspension, relates these theories to each other. Before stating it we need some definitions.

1.6 DEFINITIONS: Let $\xi \downarrow B$ be a (real) vector bundle. The *Thom-space over* B *of* ξ, or more briefly, the *ξ-sphere*, $\mathbb{S}_B^\xi$, is defined as the fiber-quotient $D(\xi)/_B S(\xi)$ of the associated disk-bundle divided by the associated sphere bundle, $(x \sim y$ iff $x, y \in S_b(\xi)$ for some $b \in B$, where $S_b(\xi)$ denotes the spherical fibre of $S(\xi)$ over $b)$.

1.7 DEFINITION (James): An *ex-space* of B is a space over B, $q: X \to B$ provided with a section $s: B \to X$ $(qs = id_B)$.

$\mathbb{S}_B^\xi$ has a natural ex-space structure.

We shall say that an ex-space $B \xrightarrow{s} X \xrightarrow{q} B$ belongs to the category $\mathfrak{W}_B^2$ if the pair (X, sB) is an object there of.

1.8 DEFINITIONS: Let $X \to B$ be an ex-space and $\xi \downarrow B$ be a vector bundle. We define the *ξ-suspension* of X by

$$\Sigma_B^\xi X := \mathbb{S}_B^\xi \wedge_B X$$

where $\wedge_B$ means fiberwise smash product. It has again the structure of an ex-space in a natural way.

In what follows we shall assume that $\mathfrak{W}_B^2$ is closed under ξ-suspensions for every ξ.

The next axiom is

1.9 SUSPENSION: For every vector bundle $\xi \downarrow B$ and every ex-space $X \to B$ there is a natural isomorphism

$$\sigma^{\xi} : \tilde{h}_B^b(X) \to \tilde{h}_B^{b+[\xi]}(\Sigma_B^{\xi} X)$$

where $[\xi] \in KO(B)$ is the class of ξ and $\tilde{h}_B^b(X) := h_B^b(X, sB)$. Equivalently one has an isomorphism

$$\sigma^{\xi} : h_B^b(X,A) \to h_B^{b+[\xi]}((\xi, \xi - 0) \underset{B}{\times} (X,A))$$

1.10 Every $b \in KO(B)$ has the form $b = k - [\xi]$, i.e. $b + [\xi] = k$, $k \in \mathbb{Z}$. Whence, by 1.9,

1.11 $$\tilde{h}_B^b = \tilde{h}_B^k \circ \Sigma_B^{\xi}$$

This means that the $KO(B)$-graded cohomology is totally determined by its $\mathbb{Z}$-graded part. On the other hand, given a $\mathbb{Z}$-graded cohomology (over B), using 1.11 as definition, one can define the associated $KO(B)$-graded cohomology. What is then the interest of studying $KO(B)$-graded theories? As we said in the introduction, the $KO(B)$-graded theory appears in a natural manner; also the relation between $RO(G)$-graded G-cohomology and $KO(B)$-graded cohomology over B illustrated in paragraph 3 justifies its study.

1.12 We now relate cohomologies parametrized by different spaces, that is, we give the so called *change-of-base* structure. To that end we pass to the more general concept of a KO-*graded parametrized cohomology theory*. This is a family $h = \{h_B\}$ of cohomologies over B, one for each compact space B, such that for each continuous mapping $\beta : B' \to B$ between compact spaces and for each object $A \subset X \to B$ in $\mathfrak{W}_B^2$ for which the pull-back $\beta^*A \subset \beta^*X \to B'$ lies in the corresponding category $\mathfrak{W}_{B'}^2$, there is a homomorphism

1.13 $$\beta^* : h_B^b(X,A) \to h_{B'}^{\beta^*(b)}(\beta^*X, \beta^*A)$$

for every $b \in KO(B)$ satisfying the following two axioms.

1.14 FUNCTORIALITY:
(i) If $\alpha : B'' \to B'$ and $\beta : B' \to B$ are continuous mappings between compact spaces, then

$$(\beta\alpha)^* = \alpha^*\beta^* : h_B^b(X,A) \to h_{B''}^{\alpha^*\beta^*(b)}(\alpha^*\beta^*X, \alpha^*\beta^*A)$$

and

(ii) $$(id_B)^* = id : h_B^b(X,A) \to h_B^b(X,A)$$

1.15 NATURALITY: If $\varphi : (X',A') \to (X,A)$ is a morphism in $\mathfrak{W}_B^2$ such that its pull-back $\beta^*\varphi : (\beta^*X', \beta^*A') \to (\beta^*X, \beta^*A)$ is a morphism in $\mathfrak{W}_{B'}^2$, then the diagram

$$\begin{array}{ccc} h_B^b(X,A) & \xrightarrow{\beta^*} & h_{B'}^{\beta^*(b)}(\beta^*X,\ \beta^*A) \\ {\scriptstyle\varphi^*}\downarrow & & \downarrow{\scriptstyle(\beta^*\varphi)^*} \\ h_B^b(X',A') & \xrightarrow{\beta^*} & h_{B'}^{\beta^*(b)}(\beta^*X',\ \beta^*A') \end{array}$$

for ever $b \in KO(B)$.

1.16 One can speak about *multiplicativity*. A cohomology over B is *multiplicative* if there is a *cup-product*

$$\cup : h_B^b(X,A) \otimes h_B^{b'}(X,A') \to h_B^{b+b'}(X,A \cup A')$$

or, equivalently, a *cross-product*

$$\times : h_B^b(X,A) \otimes h_B^{b'}(X',A') \to h^{b+b'}(X \underset{B}{\times} X',\ A \underset{B}{\times} X' \cup X \underset{B}{\times} A')$$

satisfying the usual conditions.

2. THE THEORY FIX

2.1 Let B be compact and let $q: X \to B$ be continuous, where X is a metric space. Let $p: E \to X$ be an ENR_X and $\xi, \eta \downarrow B$ be vector bundles. A (ξ,η)-*fixed point situation over* X is a commutative triangle

2.2
$$\begin{array}{ccc} \eta \underset{B}{\times} E \supset V & \xrightarrow{f} & \xi \underset{B}{\times} E \\ & \searrow \quad \swarrow & \\ & X & \end{array}$$

where $\underset{B}{\times}$ means fibered product over B obtained by considering E as a space over B via qp. f is *compactly* (ξ,η)-*fixed* if its (ξ,η)-*fixed point set*

2.3
$$\mathrm{Fix}^{\xi,\eta}(f) = \{(w,e) \in V \mid f(w,e) = (0,e) \in \xi \underset{B}{\times} E\}$$

lies properly over X. We call 2.2 a *compactly* (ξ,η)-*fixed point situation over* (X,A) if furthermore $\mathrm{Fix}^{\xi,\eta}(f)$ lies over $X-A$.

We declare two compactly (ξ,η)-fixed point situations over (X,A)

$$\begin{array}{ccc} \eta \underset{B}{\times} E_0 \supset V_0 & \xrightarrow{f_0} & \xi \underset{B}{\times} E_0 \\ & \searrow \quad \swarrow & \\ & X & \end{array} \quad \text{and} \quad \begin{array}{ccc} \eta \underset{B}{\times} E_1 \supset V_1 & \xrightarrow{f_1} & \xi \underset{B}{\times} E_1 \\ & \searrow \quad \swarrow & \\ & X & \end{array}$$

equivalent if there exists a compactly (ξ,η)-fixed point situation over $(X \times I,\ A \times I)$

$$\begin{array}{ccc} \eta \underset{B}{\times} E \supset V & \xrightarrow{f} & \xi \underset{B}{\times} E \\ & \searrow \quad \swarrow & \\ & X \times I & \end{array}$$

such that its restrictions to $X \times \{0\}$ and $X \times \{1\}$ are the given situations. We denote by $\mathrm{FIX}_B^{\xi,\eta}(X,A)$ the set of equivalence classes and we give to it an addi-

tive structure by disjoint union.
We omit B if there is no confusion.

To prove that $FIX^{\xi,\eta}(X,A)$ **depends** only on the difference $[\xi]-[\eta] = (\xi,\eta) \in KO(B)$ we need certain prerequisites.

2.4 LEMMA: $FIX^{\xi,\eta}(X,A) \doteq FIX^{\xi\oplus\zeta,\eta\oplus\zeta}(X,A)$ *for an arbitrary vector bundle* $\zeta\downarrow B$.

Proof: Let

$$\varphi : FIX^{\xi,\eta}(X,A) \to FIX^{\xi\oplus\zeta,\eta\oplus\zeta}(X,A)$$

be defined by

$$[\eta \times_B E \supset V \xrightarrow{f} \xi \times_B E] \mapsto [(\eta \oplus \zeta) \times_B E \supset W \xrightarrow{\varphi(f)} (\xi \oplus \zeta) \times_B E],$$

where $\varphi(f)$ is given by the following composite

$$\begin{array}{ccccc}
(\eta \oplus \zeta) \times_B E \supset W & & \xrightarrow{\varphi(f)} & & (\xi \oplus \zeta) \times_B E \\
\| \qquad\qquad \| & & & & \| \\
\eta \times_B \zeta \times_B E \supset (1 \times_B r)^{-1}V & & & & \xi \times_B \zeta \times_B E \\
{\scriptstyle 1 \times_B r}\downarrow \qquad\qquad \downarrow & & & & \uparrow {\scriptstyle 1 \times_B i} \\
\eta \times_B E \supset V & & \xrightarrow{f} & & \xi \times_B E
\end{array}$$

that is, $\varphi[f] = [(1 \times_B i)f(1 \times_B r)]$, where $i(e) = (0,e)$ and $r(z,e) = e$.

Conversely let

$$\psi: FIX^{\xi\oplus\zeta,\eta\oplus\zeta}(X,A) \to FIX^{\xi,\eta}(X,A)$$

be given by

$$[(\eta \oplus \zeta) \times_B E \supset W \xrightarrow{g} (\xi \oplus \zeta) \times_B E)] \mapsto [\eta \times_B (\zeta \times_B E) \supset W \xrightarrow{g} \xi \times_B (\zeta \times_B E)]$$

that is, $\psi[g] = [g]$, g viewed as a (ξ,η)-situation. Now we prove

(a) $\psi\varphi = id$:

$$\psi\varphi[f] = [(1 \times_B i)f(1 \times_B r)] = [f]$$

by lemma 2.5 that follows.

(b) $\varphi\psi = id$:

$$\varphi\psi[g] = \varphi[g] = [(1 \times_B i)g(1 \times_B r)]$$

in a diagram

$$\begin{array}{ccccc}
(\eta \oplus \zeta) \times_B (\zeta \times_B E) \supset W' & & \xrightarrow{\varphi\psi(g)} & & (\xi \oplus \zeta) \times_B (\zeta \times_B E) \\
\| & \| & & & \| \\
\eta \times_B (\xi \times_B \zeta \times_B E) \supset (1 \times_B r)^{-1}W & & & & \xi \times_B (\zeta \times_B \zeta \times_B E) \\
{\scriptstyle 1 \times_B r}\downarrow & \downarrow & & & \uparrow{\scriptstyle 1 \times_B i} \\
\eta \times_B (\zeta \times_B E) & & & & \xi \times_B (\zeta \times_B E) \\
\| & & & & \| \\
(\eta \oplus \zeta) \times_B E & \supset \bar{W} & \xrightarrow{g} & & (\xi \oplus \zeta) \times_B E
\end{array}$$

Again by lemma 2.5 we have $\varphi\psi[g] = [g]$. □

2.5 LEMMA: *Let* $i: E \to D$ *and* $r: D \to E$ *be maps over* X *between* ENR_X*'s, such that* $ri = id_E$. *If* $\eta \times_B E \supset V \xrightarrow{f} \xi \times_B E$ *is compactly* (ξ,η)*-fixed over* (X,A), *then so is*

$$\eta \times_B D \supset (1 \times_B r)^{-1}V \xrightarrow{f' = (1 \times_B i)\, f\, (1 \times_B r)} \xi \times_B D$$

and both represent the same element in $FIX^{\xi,\eta}(X,A)$.

The proof of this lemma is analogous to that of [Do 1, 4.7], whence we omit it. □

2.6 THEOREM: $FIX^{\xi,\eta}(X,A)$ *depends only on the difference* $b = [\xi]-[\eta] = (\xi,\eta) \in KO(B)$. *Thus henceforth we shall denote* $FIX^{\xi,\eta}(X,A)$ *simply by* $FIX^b(X,A)$.

Proof: $[\xi]-[\eta] = [\xi']-[\eta']$ iff there exist ζ such that

2.6 $$\xi \oplus \eta' \oplus \zeta \cong \xi' \oplus \eta \oplus \zeta$$

By lemma 2.4 we have

$$FIX^{\xi,\eta}(X,A) \cong FIX^{\xi\oplus\eta'\oplus\xi,\,\eta\oplus\eta'\oplus\xi}(X,A)$$
$$\wr\|$$
$$FIX^{\xi',\eta'}(X,A) = FIX^{\xi'\oplus\eta\oplus\zeta,\,\eta'\oplus\eta\oplus\zeta}(X,A)$$ □

2.7 REMARK: If $b = k = [\epsilon^k] \in KO(B)$, $FIX^b(X,A) = FIX^k(X,A)$ as defined in [Pr] and analogously if $b = -k = -[\epsilon^k]$.

2.8 Given an element in $FIX^b(X,A)$ we can apply to it the "forgetful" homomorphism that forgets A to get an element in $FIX^b(X)$ that obviously restricts to the trivial element in $FIX^b(A)$. As in [Pr, 4.3] this shows that the composite

2.9 $$FIX^b(X,A) \to FIX^b(X) \to FIX^b(A)$$

is zero. In fact, in an analogous way as [Pr, 4.4] we have

2.10 THEOREM: *If* $A \subset X$ *is closed, then the sequence* 2.9 *is exact.* □

Obviously we have also the

2.11 THEOREM: *If* $\alpha_0, \alpha_1\colon (X',A') \to (X,A)$ *are fiberwise homotopic maps of pairs, then* $\alpha_0^* = \alpha_1^*\colon \mathrm{FIX}^b(X,A) \to \mathrm{FIX}^b(X',A')$. □

And as in [Pr, 4.7], the

2.12 COROLLARY: *If the pair* (X,A) *is fiberwise homotopically equivalent to a pair* (X',A') *with* $A' \subset X'$ *closed, then* 2.9 *is also exact in this case.* □

We define the suspension isomorphism as follows. Let $\zeta \downarrow B$ be a vector bundle. We define the isomorphism in terms of the suspension given by multiplying over B by $(\zeta,\zeta-0)$, or equivalently, by $(\zeta,\zeta-\mathring{D}(\zeta))$ where $\mathring{D}(\zeta)$ is the associated open-disk-bundle. We define

$$\sigma^\zeta\colon \mathrm{FIX}^b(X,A) \to \mathrm{FIX}^{b+[\zeta]}((\zeta,\zeta-\mathring{D}(\zeta)) \underset{B}{\times} (X,A)) \tag{2.13}$$

by sending $\eta \underset{B}{\times} E \supset V \xrightarrow{f} \xi \underset{B}{\times} E$ (over (X,A)) to

$$\sigma^\zeta(f)\colon W \to (\zeta \oplus \xi) \underset{B}{\times} (\zeta \underset{B}{\times} E)$$

where $W = \{(y,z,e) \in \eta \underset{B}{\times} (\zeta \underset{B}{\times} E) \mid (y,e) \in V\}$ and $\sigma^\zeta(f)(y,z,e) =$
$= (-z, f'(y,e), z, f_2(y,e))$ if $f(y,e) = f'(y,e), f_2(y,e)) \in \xi \underset{B}{\times} E$.

2.14 THEOREM: σ^ζ *is an isomorphism for every* $\zeta\downarrow B$.

Proof: In order to avoid unnecesary complications we asume that $A = \phi$. The general case can be proved in exactly the same way being careful to see where the associated fixed point sets lie. We construct an inverse

$$\sigma'\colon \mathrm{FIX}^{b+[\zeta]}((\zeta,\zeta-\mathring{D}(\zeta)) \underset{B}{\times} X) \to \mathrm{FIX}^b(X)$$

as follows. Take an element in $\mathrm{FIX}^{b+[\zeta]}((\zeta,\zeta-\mathring{D}(\zeta)) \underset{B}{\times} X)$; it has a representative of the form

$$\begin{array}{ccc} \eta \underset{B}{\times} [\mathbb{R}^n \times (\zeta \underset{B}{\times} X)] \supset V & \xrightarrow{g} & (\xi \oplus \zeta) \underset{B}{\times} [\mathbb{R}^n \times (\zeta \underset{B}{\times} X)] \\ & \searrow \quad \swarrow & \\ & \zeta \underset{B}{\times} X & \end{array} \tag{2.15}$$

where $\mathrm{Fix}^{\xi\oplus\zeta,\eta}(g) \subset \eta \underset{B}{\times} [\mathbb{R}^n \times (\mathring{D}(\xi) \underset{B}{\times} X)]$. Let $(g', g_0, \bar{g}, z, x)$ be the components of g. Define

$$\begin{array}{ccc} \eta \underset{B}{\times} [(\mathbb{R}^n \times \zeta) \underset{B}{\times} X] \supset V & \xrightarrow{\sigma'(g)} & \xi \underset{B}{\times} [(\mathbb{R}^n \times \zeta) \underset{B}{\times} X] \\ & \searrow \quad \swarrow & \\ & X & \end{array}$$

by $\sigma'(g)(y,v,z,x) = (g'(y,v,z,x), \bar{g}(y,v,z,x), z + g_0(y,v,z,x), x)$. This map is compactly (ξ,η)-fixed because

$$\mathrm{Fix}^{\xi,\eta}(\sigma'(g)) = \{(y,v,z,x) \in V \mid g'(y,v,z,x) = 0,\ \bar{g}(y,v,z,x) = v,$$
$$z + g_0(y,v,z,x) = z\} = \mathrm{Fix}^{\xi\oplus\zeta,\eta}(g)$$

which lies properly over X as one can see in the diagram

$$\begin{array}{ccc} \mathrm{Fix}^{\xi,\eta}(\sigma'(g)) & = & \mathrm{Fix}^{\xi\oplus\zeta,\eta}(g) & \\ \downarrow & & \downarrow & \searrow \\ X & \xleftarrow{\text{proj}} & \zeta \underset{B}{\times} X & \supset D(\zeta) \underset{B}{\times} X. \end{array}$$

To be able to prove that σ' determines a well defined inverse of σ one proceeds in a totally analogous way to the corresponding part of the proof of 4.9 in [Pr]. □

We have, in particular, as in op.cit. 4.15, an isomorphism

2.16 $$\sigma\colon \mathrm{FIX}^b(A) \to \mathrm{FIX}^{b+1}((I,\partial I) \times A)$$

so that in order to define

2.17 $$\delta\colon \mathrm{FIX}^b(A) \to \mathrm{FIX}^{b+1}(X,A)$$

it is enough to define

2.18 $$\delta'\colon \mathrm{FIX}^{b+1}((I,\partial I) \times A) \to \mathrm{FIX}^{b+1}(X,A)$$

which can be done in an anlogous way to op.cit, and one has the

2.19 THEOREM: *The sequence*

$$\cdots \to \mathrm{FIX}^b(X) \xrightarrow{i^*} \mathrm{FIX}^b(A) \xrightarrow{\delta} \mathrm{FIX}^{b+1}(X,A) \xrightarrow{j^*} \mathrm{FIX}^{b+1}(X) \to \cdots$$

is exact, where i *and* j *are inclusions.* □

2.20 PROPOSITION: *Let* $A \subset X$ *be closed and assume that* C *is such that its closure* $\bar{C}$ *is contained in the interior* $\mathring{A}$ *of* A; *then the canonical homomorphism*

$$\mathrm{FIX}^b(X,A) \to \mathrm{FIX}^b(X-C,\ A-C)$$

is an isomorphism.

The *proof* is analogous to that of 4.26 in [Pr]. □

With this last proposition we have the

2.21 THEOREM: *The functors* FIX^b, $b \in KO(B)$ *together with the natural transformations* δ *and* σ^ζ *constitute a multiplicative* KO-*graded parametrized cohomology theory in the category of pairs of metric spaces over* B, *for compact spaces* B.

Proof. The homotopy axiom is 2.11; the excision axiom is 2.20; the exactness axiom is 2.14.

The multiplicative structure is a given as follows. Define

$$\cup\colon \mathrm{FIX}^b(X,A) \otimes \mathrm{FIX}^{b'}(X,A') \to \mathrm{FIX}^{b+b'}(X,A \cup A')$$

in representatives

$$\begin{array}{ccc} \eta \underset{B}{\times} E \supset V \xrightarrow{f} \xi \underset{B}{\times} E & & \eta' \underset{B}{\times} E' \supset V' \xrightarrow{f'} \xi' \underset{B}{\times} E' \\ \searrow \quad \swarrow & & \searrow \quad \swarrow \\ (X,A) & & (X,A') \end{array}$$

by representing $[f] \cup [f']$ by

$$\begin{array}{c} (\eta\oplus\eta')\underset{B}{\times}(E\underset{B}{\times}E') \approx (\eta\underset{B}{\times}E)\underset{B}{\times}(\eta'\underset{B}{\times}E') \supset V\underset{B}{\times}V' \xrightarrow{f\times f'} (\xi\underset{B}{\times}E)\underset{B}{\times}(\xi'\underset{B}{\times}E') \approx (\xi\oplus\xi')\underset{B}{\times}(E\underset{B}{\times}E') \\ \searrow \qquad \swarrow \\ (X, A \cup A') \end{array}$$

The <u>change-of-base structure</u> is given in the following manner:

Let $\beta : B' \to B$ be continuous and $A \subset X \to B$ be a pair of metric spaces over B.

$$\beta^* : FIX_B^b(X,A) \to FIX_{B'}^{\beta^*(b)}(\beta^*X, \beta^*A)$$

is given in terms of representatives by associating to

$$\begin{array}{c} \eta \underset{B}{\times} E \supset V \xrightarrow{f} \xi \underset{B}{\times} E \\ \searrow \quad \swarrow \\ (X,A) \end{array}$$

The representative of $\beta^*[f]$ which is the pull-back over β

$$\begin{array}{c} \beta^*\eta \underset{B'}{\times} \beta^*E \supset \beta^*V \xrightarrow{\beta^*f} \beta^*\xi \underset{B'}{\times} \beta^*E \\ \searrow \quad \swarrow \\ (\beta^*X, \beta^*A) \end{array}$$

□

2.22 REMARK: According to 1.10, given any element $b \in KO(B)$ there exists $\xi \downarrow B$ such that $b + [\xi] = k \in KO(B)$ so that by suspending we have

2.23 $$FIX^b(X,A) \cong FIX^k((\xi, \xi - 0) \underset{B}{\times} (X,A))$$

but the right hand side is isomorphic to $\pi_s^k((\xi, \xi - 0) \underset{B}{\times} (X,A))$ by 4.27 (iii) in [Pr]. This, by formula 1.11, is by definition $\pi_B^b(X,A)$, $KO(B)$-*graded stable cohomotopy over* B. Whence

2.24 THEOREM: $FIX^b \triangleq \pi_B^b$ *for all* $b \in KO(B)$. □

§3. EQUIVARIANT FIX AND ITS RELATION TO PARAMETRIZED FIX.

In this paragraph I shall give in a very concise way the axioms that characterize the $RO(G)$-graded G-cohomology theories and I shall illustrate them with the example G-FIX which constitutes a theory isomorphic to G-cohomotopy, as defined, say, by tom Dieck in [tD] or Kosniowski in [Ko]. The theory G-FIX has also been partly studied by Ulrich in [Ul]. I am grateful to him for fruitful discussions about equivariant fixed point theory.

To conclude the section I shall give a natural transformation between G-FIX and

FIX_{B^nG} where $\ldots \subset B^nG \subset B^{n+1}G \subset \ldots$ is some suitable filtration of the classifying space BG of G and I shall discuss its relation to the (already proved) Graeme Segal conjecture about the Burnside ring A(G).

3.1 Let G be a compact Lie Group and let $G\text{-}\mathfrak{W}^2$ be a category of pairs of G-spaces, $A \subset X$. An RO(G)-*graded* G-*cohomology theory* (a G-*cohomology* for short consists of a family of contravariant functors

$$G\text{-}h^\rho = h^\rho : G\text{-}\mathfrak{W}^2 \to Ab$$

and a family of natural transformations

$$\delta = \delta^\rho : h^\rho \circ T \to h^{\rho+1}$$

for every $\rho \in RO(G)$ where $T(X,A) = (A,\phi) = A$, satisfying the axioms of HOMOTOPY (for G-homotopic equivariant maps f_0, f_1: $(X', A') \to (X, A)$), EXCISION (for C such that $\bar{C} \subset \overset{\circ}{A}$, C G-invariant), EXACTNESS:

$$\ldots \to h^\rho(X) \to h^\rho(A) \xrightarrow{\delta} h^{\rho+1}(X,A) \to h^{\rho+1}(X) \to \ldots$$

has to be exact. For the last axiom we need the

3.2 DEFINITION: Let M be a (real) G-module. The M-*dimensional* G-*sphere*, or more briefly, the M-*sphere*, $\mathbb{S}^M$, is defined as the one-point compactification of M with the canonical action, or, equivalently, as the quotient D(M)/S(M) of the unitary disk of M by the unitary sphere (since we may always assume that G acts orthonormally in $\mathbb{R}^n$ in the definition of M). ∞, resp. {S(M)} is a fixed point of the action of G.

A *pointed* G-*space* X is a G-space X with a base point x_0 which is a fixed point. $\mathbb{S}^M$ is a pointed G-space.

3.3 DEFINITION: The M-*suspension* of a pointed G-space X is defined by

$$G\text{-}\Sigma^M X = \Sigma^M X := \mathbb{S}^M \wedge X,$$

where $\wedge$ is the usual smash product and $\mathbb{S}^M \wedge X$ has the diagonal G-action. So we have also the axiom of

3.4 SUSPENSION: For every irreducible G-module M and every pointed G-space there is a natural isomorphism

$$\sigma^M : \tilde{h}^\rho(X) \to \tilde{h}^{\rho+[M]}(\Sigma^M X)$$

where $[M] \in RO(G)$ is the class of M and $\tilde{h}^\rho(X) := h^\rho(X, x_0)$. Equivalently, there is an isomorphism

$$\sigma^M : h^\rho(X,A) \to h^{\rho+[M]}((M, M-0) \times (X,A))$$

3.5 (cf. 1.10) The RO(G)-graded case differs from the KO(B)-graded case in

That a G-module N need not have an "inverse", i.e. a G-module M such that M N is equivalent to the trivial representation. This makes the G-cohomologies <u>essentially distinct</u> from those graded by $\mathbb{Z}$.

3.6 As for KO-graded parametrized cohomology, there is here a concept of RO-*graded equivariant cohomology theory* (RO is the functor $G \mapsto RO(G)$). This is a family $h = \{G\text{-}h\}$ of G-cohomologies, one for each compact Lie group G (or at least for each group in some suitable category of topological groups, e.g. finite groups), such that for each homomorphism $\varphi\colon G' \to G$ and for each object $(X,A) \in G\text{-}\mathfrak{W}^2$ for which the pair (X,A) with the action of G' induced by φ lies in $G'\text{-}\mathfrak{W}^2$, there is a homomorphism

$$\varphi^*\colon G\text{-}h^{\rho}(X,A) \to G'\text{-}h^{\varphi^*(\rho)}(X,A)$$

for every $\rho \in RO(G)$ satisfying corresponding conditions of FUNCTORIALITY and NATURALITY. This is the *change-of-group*-structure. It has to satisfy as well the fact that the composite

$$G\text{-}h^{\rho}(G/H) \to H\text{-}h^{\rho|H}(G/H) \to H\text{-}h^{\rho|H}(*)$$

is an isomorphism for all (closed) $H \subset G$ (cf. [Ko] . The second map is induced by $H/H \subset G/H$.).

3.7 There is also a concept of *multiplicativity*.

3.8 Let G be a compact Lie group and X a metric G-space. Let $p\colon E \to X$ be a $G\text{-}ENR_X$ (i.e. E is a G-retract over X of an invariant neighborhood $V \subset M \times X$, where M is a G-module and $M \times X$ has the diagonal action of G) and let M,N be real G-modules. An (M,N)-*fixed point* G-*situation over* X is an equivariant commutative triangle

3.9
$$\begin{array}{ccc} N \times E \supset V & \xrightarrow{\;f\;} & M \times E \\ & \searrow \quad \swarrow & \\ & X & \end{array}$$

f is *compactly* (M,N)-*fixed* if its (M,N)-*fixed point set*

3.10
$$\mathrm{Fix}^{M,N}(f) = \{(n,e) \in V \mid f(n,e) = (0,e) \in M \times E\}$$

lies properly over X. We say that 3.9 is a G-situation over (X,A) if furthermore $\mathrm{Fix}^{M,N}(f)$ lies over X-A. Two such situations over (X,A) are *equivalent* if they are restrictions to $X \times \{0\}$ and $X \times \{1\}$ of such a situation over $(X \times I, A \times I)$ (considering $X \times I$ and $A \times I$ as G-spaces with the trivial action on I). We denote by $G\text{-}FIX^{M,N}(X,A)$ the set of equivalence classes. We have

3.11 THEOREM: $G\text{-}FIX^{M,N}(X,A)$ *depends only on the difference* $\rho = [M] - [N] =$

$= (M,N) \in RO(G)$. *Thus we denote it henceforth by* $G\text{-}FIX^{\rho}(X,A)$ *or simply by* $FIX^{\rho}(X,A)$ *if this does not lead to confusion.*

The *proof* is essentialy the same as that of 2.6. □

3.12 G-FIX has a structure of an RO(G)-graded G-cohomology theory. The proof is essentially the same as that given in §4 in [Pr]. We should observe that Tietze's lemma has an equivariant version, Tietze-Gleason's theorem [Br, I.2.3], which allows us to write the proofs of [Pr] in the equivariant case.

It is, however, convenient to give an explicit definition of the suspension isomorphism. We give it in terms of the suspension given by multiplying by $(P,P-0)$ or, equivalently, by $(P, P-\mathring{D}(P))$, where $\mathring{D}(P)$ is the open unit disk if the G-module P is orthogonal. For P we define

3.13 $$\sigma^P: FIX^{\rho}(X,A) \to FIX^{\rho+[P]}((P, P-\mathring{D}(P)) \times (X,A))$$

by sending $N \times E \supset V \xrightarrow{f} M \times E$ over (X,A) to

$$\sigma^P(f): W \to (P \oplus M) \times (P \times E)$$

where $W = \{(n,p,e) \in N \times (P \times E) | (n,e) \in V\}$ and $\sigma^P(f)(n,p,e) = (-p, f'(n,e), p, f_2(n,e))$ if $f(n,e) = (f'(n,e), f_2(n,e)) \in M \times E$.

With an analogous proof as that of 2.14 one has

3.14 THEOREM: σ^P *is an isomorphism for every* G-*module* P. □

Here after we shall assume that all G-modules considered are orthogonal.

3.15 DEFINITION. If M,N are G-modules, we define the stable (M,N)-*cohomotopy group* of the G-space X as

$$G\text{-}\pi^{M,N}(X) := \operatorname*{colim}_P [\Sigma^P\Sigma^N (X^+), \Sigma^P S^M]_G$$

that is, the stabilization of the group of G-homotopy classes of G-maps of the N-suspension of $X^+ := X \cup \{\infty\}$ into the M-sphere (3.2), where P varies in the category (made small) of all unitary (complex) G-modules, the direction given by $P \leqslant Q \Leftrightarrow$ there exists R such that $P \oplus R = Q$

Proceeding in analogous manner to 3.2 in [Pr] or to the proof of 4.3 in [Do] it is possible to prove the

3.16 PROPOSITION: $G\text{-}FIX^{M,N}(X) \cong G\text{-}\pi^{M,N}(X)$ *for every pair of orthogonal representations* M,N.

3.17 COROLLARY: $G\text{-}\pi^{M,N}(X)$ *depends only on the difference* $[M] - [N] \in RO(G)$.

In summary we have

3.18 THEOREM: (cf. [U1], 4.13-14). (i) *The functors* $G\text{-}FIX^{\rho}$ *constitute an* $RO(G)$*-graded multiplicative* G*-cohomology theory in the category of metric* G*-spaces,*
(ii) *one has an index homomorphism* $I^{\rho}: G\text{-}FIX^{\rho} \to h^{\rho}$ *for every* $RO(G)$*-graded multiplicative* G*-cohomology theory* h, *and*
(iii) $I^{\rho}: G\text{-}FIX^{\rho} \to G\text{-}\pi^{\rho}$ *is a natural equivalence of* G*-cohomology theories.* □

3.19 I shall describe now the relation between G-FIX and FIX_{BG}. To begin let $P \to B$ be a principal G-bundle over a compact space B. There is a *Segal-homomorphism*

$$s: G\text{-}FIX^{\rho}(X,A) \to FIX_B^{a(\rho)}(P \underset{G}{\times} X, P \underset{G}{\times} A)$$

where $a: RO(G) \to KO(B)$ is Atiyah's map sending $[M]-[N]$ to $[P \underset{G}{\times} M] - [P \underset{G}{\times} N]$. s is given as follows: An element in $G\text{-}FIX^{\rho}(X,A)$ represented by

3.20
$$\begin{array}{ccc} N \times E \supset V & \xrightarrow{f} & M \times E \\ & \searrow \;\; \swarrow & \\ & X & \end{array}$$

is sent to the element in $FIX_B^{a(\rho)}(P \underset{G}{\times} X, P \underset{G}{\times} A)$ represented by

$$\begin{array}{ccc} (P\underset{G}{\times}N) \underset{B}{\times} (P\underset{G}{\times}E) = P\underset{G}{\times}(N\times E) \supset P\underset{G}{\times}V & \xrightarrow{1\underset{G}{\times}f} & P\underset{G}{\times}(M\times E) = (P\underset{G}{\times}M) \underset{B}{\times} (P\underset{G}{\times}E) \\ & \searrow \;\; \swarrow & \\ & P \underset{G}{\times} X & \end{array}$$

Hence forth we shall be interested in the bundles $E^nG \to B^nG$ of Milnor's filtration (or any others given by compact B^nG) of the universal bundle $EG \to BG$. We restrict ourselves to the case of G <u>a finite group.</u>

3.21 The groups $G\text{-}FIX^{\rho}(X,A)$ are modules over the Burnside ring $A(G)$ in an obvious way. If $\tilde{A}(G)$ denotes the augmentation ideal in $A(G)$ we can complete $G\text{-}FIX^{\rho}(X,A)$ as

$$G\text{-}FIX^{\rho}(X,A)^{\wedge} = \lim_{n} (G\text{-}FIX^{\rho}(X,A)/(\tilde{A}(G)^{n+1}))$$

where $(\tilde{A}(G)^{n+1})$ denotes the subideal of the elements divisible by products of $n+1$ elements in $\tilde{A}(G)$.

3.22 Let $X \to BG$ be a (metric) space over BG and let $b \in KO(BG) = \lim_{n} KO(B^nG)$ be represented by $b^n \in KO(B^nG)$ for every n. By the change-of-base structure there are restriction homomorphisms

$$FIX_{B^{n+1}G}^{b^{n+1}}(X^{n+1}, A^{n+1}) \to FIX_{B^nG}^{b^n}(X^n, A^n)$$

where (X^n, A^n) is the restriction of (X,A) to B^nG. Define

$$\mathrm{FIX}^{b}_{BG}(X,A) = \lim_{n} \mathrm{FIX}^{b^n}_{B^nG}(X^n,A^n).$$

3.23 We have, as in 3.19, Segal-homomorphisms

$$s\colon G\text{-}\mathrm{FIX}^{\rho}(X,A) \to \mathrm{FIX}^{a^n(\rho)}_{B^nG}(E^nG \underset{G}{\times} (X,A))$$

it is easy to see that it factors through $G\text{-}\mathrm{FIX}^{\rho}(X,A)/(\tilde{A}(G)^{n+1})$; thus we have a homomorphism

3.24 $$s\colon G\text{-}\mathrm{FIX}^{\rho}(X,A)^{\wedge} \to \mathrm{FIX}^{a(\rho)}_{BG}(EG \underset{G}{\times} (X,A)).$$

Now, $A(G)$ is a noetherian ring, whence completion is an exact functor for finitely generated modules over $A(G)$ (see [At'], 3.16). This proves

3.25 PROPOSITION. *The family of completed functors* $G\text{-}\mathrm{FIX}^*(\ )^{\wedge}$ *is again an* $RO(G)$*-graded cohomology theory, when restricted to finite* G*-CW-pairs (or to any category of* G*-pairs* (X,A) *such that* $G\text{-}\mathrm{FIX}^{\rho}(X,A)$ *is finitely generated over* $A(G)$ *for all* $\rho \in RO(G)$. □

The homomorphism 3.24 is a natural transformation of G-cohomology theories in the sense of Kosniowski [Ko]; in order to prove that it is an isomorphism, we may apply his Comparison Theorem, op.cit. 2.14. To do that we have to check that s is an isomorphism for all *orbits* G/H, that is

3.26 $$s\colon G\text{-}\mathrm{FIX}^{\rho}(G/H)^{\wedge} \overset{\cong}{\to} \mathrm{FIX}^{a(\rho)}_{BG}(EG \underset{G}{\times} (G/H));$$

but the left hand side is equal to $H\text{-}\mathrm{FIX}^{\rho}(*)^{\wedge}$, whereas the right hand side equals $\mathrm{FIX}^{a(\rho)}_{BH}(BH)$.

Carlsson's theorem [Ca] proving Segal's conjecture asserts

3.27 $$G\text{-}\mathrm{FIX}^{\rho}(*)^{\wedge} \cong G\text{-}\mathrm{FIX}^{\rho}(EG) = \lim_{n} G\text{-}\mathrm{FIX}^{\rho}(E^nG),$$

the isomorphism induced by $EG \to *$.

So, in order to prove that 3.24 is an isomorphism, we need the

3.28 LEMMA. $G\text{-}\mathrm{FIX}^{\rho}(EG) \cong \mathrm{FIX}^{a(\rho)}_{BG}(BG)$.

Proof In fact we prove that if $P \downarrow P/G = B$ is a principal G-bundle, then $G\text{-}\mathrm{FIX}^{\rho}(P) \cong \mathrm{FIX}^{a(\rho)}_{B}(B)$. The maps

$$\varphi\colon G\text{-}\mathrm{FIX}^{\rho}(P) \rightleftarrows \mathrm{FIX}^{a(\rho)}(B)\colon \psi$$

$$[F] \longmapsto [F/G]$$

$$[\mathrm{id}_P \underset{B}{\times} \bar{f}] \longleftarrow\!\shortmid [\bar{f}]$$

are inverse to each other, as one easily checks. □

Whence 3.26 is an isomorphism for all H by 3.27 and 3.28. We have

3.29 THEOREM. *Let* G *be a finite group. The homomorphism*

3.24 $$s: G\text{-}FIX^{\rho}(X,A)^{\wedge} \to FIX^{a(\rho)}_{BG}(EG \underset{G}{\times} (X,A))$$

is an isomorphism for all finite CW-*pairs* (X,A), *(or all* G-*pairs* (X,A) *such that* $G\text{-}FIX^{\rho}(X,A)$ *is finitely generated - over* A(G) *- for all* $\rho \in RO(G)$.

§4. DIAGRAMMATIC FIX-THEORY

In this paragraph we show how to extend the functor FIX to a more general set up, namely how to define FIX for diagrams of topological spaces. We shall just state the results and all details will appear elsewhere.

4.1 DEFINITION A *quiver* Q is a finite digraph consisting of
(i) a finite set $V_Q = \{1,2,3,\ldots\}$ of *vertices*,
(ii) a finite set $A_Q = \{a,b,c,\ldots\}$ of *arrows*, and
(iii) two functions $\partial_0, \partial_1 : A \to V$, assigning to each arrow its *origin* and its *end* respectively.

We shall always assume that the digraph is connected,

4.2 *Examples*

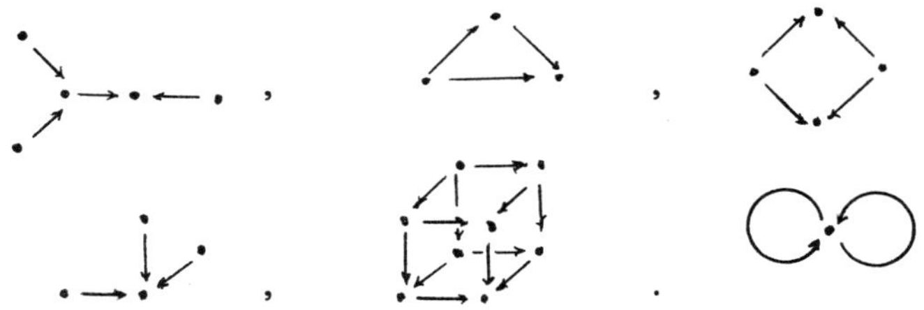

4.3 DEFINITION. A *set of relations* R on a quiver Q is a set of conditions imposed either on the category $\mathfrak{C}(Q)$ generated by Q or on the completed category $\hat{\mathfrak{C}}(Q)$ obtained from the former by including all possible pullbacks. The conditions are of the following sort:

(a) Equality of certain composites (e.g. commutativity of certain diagrams).
(b) Coincidence of certain pullbacks.

4.4. *Examples*

4.4.1 Q =  $\mathfrak{C}(Q)$ =  (+ identities)

$\hat{\mathfrak{C}}(Q)$ = 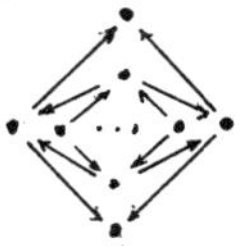 , Relations: All new vertices are the same, i.e.

$\hat{\mathfrak{C}}(Q,R)$ = 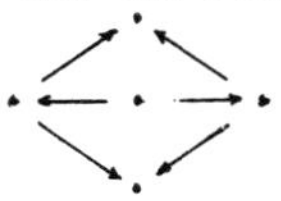

4.4.2 Q =  a, $\mathfrak{C}(Q) = \mathbf{N} = \mathfrak{C}(Q)$
Relations: $a^n = 1$, i.e. $\hat{\mathfrak{C}}(Q,R) = \mathbf{Z}/n$.

4.4.3 Any finite group G (given by generators and relations, i.e.

Q = 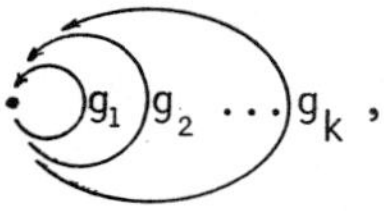

, R = given relations, i.e. $\hat{\mathfrak{C}}(Q,R) = G$.

4.4.4 Let B be a finite polyhedron. Define a quiver Q_B by taking all subpolyhedra of B as vertices and an arrow between two vertices every time the origin is contained in the end. Then Q_B is automatically provided with the relations given by the commutativity of all possible diagrams.

4.5 DEFINITION Let $\mathfrak{C}$ be a category which is closed under pullbacks. If (Q,R) is a quiver with relations, then a *$\mathfrak{C}$-representation* of (Q,R) is a functor $\underline{X}: \mathfrak{C}(Q,R) \to \mathfrak{C}$, respecting pullbacks.

Relevant *examples* of category $\mathfrak{C}$ are for us the following

$\mathfrak{C} = \mathfrak{Top}^{\mathfrak{e}}$ = (normal) topological spaces and embeddings.

We speak of *topological representations*.

$\mathfrak{C} = \mathfrak{FinSet}^{\mathfrak{e}}$ = finite sets and inclusions.

We speak of *finite representations*.

$\mathfrak{C} = \mathfrak{Vect}^{\mathfrak{e}}_{\mathbb{R}}$ = finite dimensional real vector spaces and monomorphisms

We speak of *linear* (or *euclidean*) representations.

4.6 We call a quiver with relations (Q,R) *admissible* if $\hat{\mathfrak{C}}(Q,R)$ is a finite category.

4.7 *Example*. Let B be a finite polyhedron and $p: E \to B$ be any space over B. Then p determines in a canonical way a topological representation $\underline{E}$ of Q_B by taking $\underline{E}(P) = p^{-1}(P)$ for every subpolyhedron P of B and the corresponding inclusion for every inclusion $P \subset P'$ of subpolyhedra.

4.8 From now on we shall assume that every (Q,R) considered is admissible. Already in the case of groups the theory fails if the group is not finite, e.g. for $\mathbb{Z}$, since, as noted in 3.12, a Tietze theorem is essential.

For topological representations of admissible quivers (Q,R) there is a version of Tietze's theorem, which allows us to perform the program to develop the FIX-theory.

4.9 THEOREM *There is an* RO(Q)*-graded cohomology theory* Q-FIX *defined on (normal) topological representations of* Q *for any admissible quiver* Q (with relations).

(The real *representation ring* RO(Q) of Q has an obvious definition - using $\mathfrak{Vect}^{\mathfrak{e}}_{\mathbb{R}}$.

There are two intriguing questions. The first one is related to results by Dold on the Burnside ring and the second one is related to Segal's conjecture. They read as follows

Question 1. Let A(Q) be the *Burnside ring* of Q defined in an obvious way using finite representations of Q. Is the canonical map

$$A(Q) \to Q\text{-FIX}(*)$$

an isomorphism? (cf. [Do], 3.5).

Question 2. Given a quiver (Q,R) there is a classifying space BQ (Segal's classifying space of the category $\mathbb{C}(Q,R)$) and a *principal classifying* Q-*bundle* $\underline{EQ}\downarrow BQ$ ($\underline{EQ}$ is a topological representation of Q). If $\underline{X}$ is any topological representation, there is a homomorphism

4.10 $$s: Q\text{-FIX}^*(\underline{X}) \to Q\text{-FIX}^*(\underline{X} \times \underline{EQ}).$$

Is it an isomorphism after suitably completing the left-hand-side?

This paper was partly written during a visit of the author at the Mathematisches Institut of the University of Heidelberg, to whose staff the author is grateful.

REFERENCES:

[At] Atiyah, M.F. "K-Theory", Benjamin 1967.

[At'] Atiyah, M.F. *Characters and Cohomology of Finite Groups*, Publ. Math. IHES, Nr. 9, 23-64, (1961)

[Br] Bredon, G.E. "Introduction to Compact Transformation Groups", Academic Press 1972.

[Ca] Carlsson, G. *Equivariant Stable Homotopy and Segal's Burnside Ring Conjecture*, preprint, 1982.

[tD] tom Dieck, T. "Transformation Groups and Representation Theory" LNM 766, Springer 1979.

[Do] Dold, A *The Fixed Point Index of Fibre-Preserving Maps*, Inventiones math. 25, 281-297 (1974)

[Do'] Dold, A *Fixed Point Theory and Homotopy Theory*, Contemporary Mathematics, Vol. 12, 105-115, (1982)

[Ko] Kosniowski, C. *Equivariant Cohomology and Stable Cohomotopy*, Math. Ann. 21o, 83-104 (1974)

[Pr] Prieto, C. *Coincidence Index for Fiber-Preserving Maps. An Approach to Stable Cohomotopy*, manuscripta math. 47. 233-249 (1984)

[Ul] Ulrich, H. "Der Äquivariante Fixpunktindex vertikaler G-Abbildungen", Thesis, Heidelberg. 1983

INSTITUTO DE MATEMATICAS
U.N.A.M.
04510 México, D.F.
MEXICO

Contemporary Mathematics
Volume **58**, Part II, 1987

ON THE ATIYAH-HIRZEBRUCH SPECTRAL SEQUENCE $H^*(X_n;Z/p) \Rightarrow \underleftarrow{K}^*(X_n;Z/p)$, X_n a $K(Z/p,n)$

Jack Ucci

ABSTRACT. The Atiyah-Hirzebruch spectral sequence from ordinary mod p cohomology to inverse limit mod p complex K-theory is determined for the Eilenberg-MacLane space $K(Z/p,2)$. Secondly, an algebraic spectral sequence is constructed for each $n > 2$ which is conjectured to be the above Atiyah-Hirzebruch spectral sequence for all Eilenberg-MacLane spaces $K(Z/p,n)$, $n > 2$.

0. INTRODUCTION

The complex K-theory K^*X of Eilenberg-MacLane spaces X was studied by Anderson and Hodgkin [1] and Buhštaber and Miščenko [4] in the late 1960's using the Rothenberg-Steenrod spectral sequence. One of their results states the vanishing of the inverse limit complex K-theory of most $K(G,n)$, in particular for $G = Z/p$ and $n \geq 2$.

We reexamine this result from the viewpoint of the Atiyah-Hirzebruch spectral sequence (AHss). Our considerations are restricted to the case of odd primes p. Our main result is the determination of the AHss $H^*(X_2;Z/p) \Rightarrow \underleftarrow{K}^*(X_2;Z/p)$, X_2 a $K(Z/p,2)$. The analogous AHss for X_1 a $K(Z/p,1)$ has only one nontrivial differential (d_{2p-1}) and so X_2 is the first interesting case.

Secondly, we construct algebraic elementary acyclic spectral sequences mod p of "Atiyah-Hirzebruch type". Then we inductively construct an algebraic spectral sequence $E(n)$ as a certain infinite tensor product of the proceeding elementary spectral sequences. $E(n)$ will have the correct initial and terminal data (namely, its E_{2p-1} level as a differential algebra will be isomorphic to E_{2p-1} level of the AHss $H^*(X_n;Z/p) \Rightarrow \underleftarrow{K}^*(X_n;Z/p)$) to be isomorphic as spectral sequences to this AHss. We conjecture that it is. This conjecture together with the details of our construction of $E(n)$ immediately imply that the spectral sequence converging to $\underleftarrow{K}^*(X_n;Z/p)$ associated to the filtration of $X_n \sim SP^\infty Y_n$, the infinite symmetric product of $Y_n = S^n \cup_p e^{n+1}$, by the finite symmetric products $\{SP^{p^r}Y_n\}$ $r \geq 0$ collapses at $E_2^{**} = E_\infty^{**}$ (and perhaps conversely as well). Finally the conjecture can be viewed as a possible generalization of the Cartan description of $H^*(K(Z/p,n);Z/p)$ as the cohomology of an infinite tensor product of elementary constructions.

1980 Mathematics Subject Classification. 55D20, 55F50, 55G15, 55Hxx.

Much of this paper was prepared while the author was a visitor at the Centro de Investigación del IPN during the fall 1984 semester. It is a pleasure for me to express my thanks to my hosts Enrique Antoniano and Sam Gitler for their warm hospitality and generous help.

1. $H^*(X_2;Z_{(p)})$

Consider the $H^*(\ ;Z_{(p)})$-Serre spectral sequence for the fibration $CP^\infty \to X_2 \to K(Z,3)$. Let $u_0 \in H^3(K(Z,3);Z_{(p)})$ be the fundamental class and let $x \in H^2(CP^\infty;Z_{(p)})$ be a polynomial generator of $H^*(CP^\infty;Z_{(p)}) \cong Z_{(p)}[x]$.

PROPOSITION 1.1. (a) d_3 is the only nontrivial differential in this spectral sequence and it is given (up to sign) by

$$d_3(1 \otimes x^k) = pk(u_0 \otimes x^{k-1})$$

(b) The classes $u_0 \otimes x^{k-1}$ at the E_2 level define elements at $E_4 = E_\infty$ of order p^{r+1} where $k = p^r k_1$ and $(k_1,p) = 1$. All other classes $y \otimes x^k$ at the E_2 level with $\dim y > 3$ survive to E_∞ and define elments of order p.

(c) The kernel of the mod p reduction homomorphism $r_p: H^*(X_2;Z_{(p)}) \to H^*(X_2;Z/p)$ is additively generated by the classes $p(u_0 \times x^{k-1})$, $k \geq 1$ and $(k,p) = p$.

Proof. Since $d_3(1 \otimes x) = p(u_0 \otimes 1)$, the formula in (a) follows from the multiplicative properties of the spectral sequence. Collapse at E_4 will follow from the second statement of (b) that all classes $y \times x^k$ at the E_2 level with $\dim y > 3$ survive to E_∞, since then all differentials d_r, $r > 3$, on classes of the form $Nu_0 \times x^k$ must be zero.

Observe that the $H^*(\ ;Z/p)$-Serre spectral sequence for this fibration collapses at $E_2 = E_\infty$. Also since $H^*(K(Z,3);Z)$ has no elements of order p^2 [5, Expose 11], $H^i(K(Z,3);Z_{(p)})$ is a Z/p-vector space for $i > 3$. Hence at the E_2 level, the mod p reduction homomorphism of spectral sequences is a monomorphism on all $E_2^{i,j}$ when $i > 3$. It follows that no differential in the $H^*(\ ;Z_{(p)})$-Serre spectral sequence can have a nonzero image in some $E_r^{i,j}$ when $i > 3$. This proves (a) and (b). (c) is then immediate.

REMARK 1.2. $\nu_\ell = \beta_{(p)} P^{p^{\ell-1}} P^{p^{\ell-2}} \cdots P^p P^1 u_0 = \beta_{(p)} P^{I_\ell} u_0$, $I_\ell = (p^{\ell-1}, p^{\ell-2}, \ldots, p, 1)$, define nonzero elements of $H^*(K(Z,3);Z_{(p)})$ and $\nu_\ell \otimes x^k$ are E_2 level representatives of classes (which we write as) $x^k \nu_\ell$ of order p in $H^*(X_2;Z_{(p)})$. Note that $x^k \nu_\ell$ is not a product despite its notation. In fact $x^k \nu_\ell = \beta_{(p)}(x^k u_\ell)$, $u_\ell = P^{I_\ell} u_0$, and $x^k u_\ell$ is a product of the mod p cohomology classes x^k and u_ℓ.

We need one further observation about $H^*(X_2;Z_{(p)})$. In [8,¶1] we gave a description of the cohomology ring $H^*(K(Z,3);Z_{(p)})$ in terms of generators and relations. In particular, $H^*(K(Z,3);Z_{(p)})$ is a ring with generators u_0, ν_j $(j \geq 1)$ and certain additional classes $u_{(I)} = \beta_{(p)}(u_{i_1}u_{i_2}\cdots u_{i_\ell})$ where the sequence $I = (i_1,i_2,\ldots,i_\ell)$ must satisfy some condition, and relations which include 1) $p\nu_j = 0 = pu_{(I)}$, (2) an expression for a certain sum of terms $u_{(I)}\nu_j$ and (3) an expression for $u_{(I)}u_{(J)}$. Moreover, the Z/p vector space $\sum_{i>3} H^i(K(Z,3);Z_{(p)})$ has a basis $\{u_0^t u_{(I)}^{t'} \nu_J^N\}$, $t,t' = 0$ or 1 and I, J, N are certain sequences of integers.

The 2nd stage of the Milnor construction M_2 of $K(Z,3)$ is homeomorphic to an adjunction space of the form $\Sigma CP^\infty \cup C(CP^\infty * CP^\infty)$. Moreover, M_2 can be realized as a subset of a suitable $K(Z,3)$, so we have an inclusion $i: M_2 \to K(Z,3)$. In 8 we studied the kernel and cokernel of the coboundary homomorphism $\delta: H^*(\Sigma CP^\infty;Z_{(p)}) \to H^*(M_2,\Sigma CP^\infty;Z_{(p)})$ and the result obtained in [8] easily implies

PROPOSITION 1.3. The induced homomorphism i^* on $H^*(\ ;Z_{(p)})$ of the inclusion $i: M_2 \to K(Z,3)$ annihilates all classes $u_0^t u_{(I)}^{t'} \nu_J^N$ except the idecomposables u_0, ν_j. Moreover, the p-torsion in $H^*(M_2;Z_{(p)})$ consists of copies of Z/p with generators the classes $i^*\nu_j$ in dimensions $2p^j+2$.

$H^*(M_2;Z_{(p)})$ has a large free summand as well. For example i^*u_0 is a generator of $H^3(M_2;Z_{(p)})$ while all other free generators lie in even dimensions. It follows from 1.3 that the homomrphism

$$H^*(M_2;Z_{(p)}) \xrightarrow{i^*u_0} H^*(M_2;Z_{(p)})$$

is the zero homomorphism. Hence the $H^*(\ ;Z_{(p)})$-Serre spectral sequence for the induced fibration $CP^\infty \to \pi^{-1}M_2 \to M_2$ also has only one nontrivial differential and it is completely given by $d_3(1 \times x^k) = kp(i^*u_0 \times x^{k-1})$. In particular

PROPOSITION 1.4. In the $H^*(\ ;Z_{(p)})$-Serre spectral sequence of the induced fibration $CP^\infty \to \pi^{-1}M_2 \to M_2$, all E_2 level classes of the form $i^*u_0 \otimes x^k$ or $i^*\nu_j \otimes x^k$, $k \geq 0$, survive to E_∞ to define nonzero (torsion) cohomology classes which we denote $x^k i^*u_0$, $x^k i^*\nu_j$.

2. THE AHSS FOR X_2

From Cartan [5] we have $H^*(X_2;Z/p) \cong E \otimes P$ where E is a Z/p-exterior algebra on the generators $\beta\iota_2, P^1\beta\iota_2, \ldots, P^{I_k}\beta\iota_2, \ldots$, and P is the Z/p-polynomial algebra on the generators, $\iota_2, \beta P^1\beta\iota_2, \ldots, \beta P^{I_k}\beta\iota_2, \ldots$, where $P^{I_k} = pP^{k-1}pP^{k-2}\ldots P^1$, $k \geq 1$. Rename these generators $x = \iota_2$, $\beta x = \beta\iota_2$, $e_k = P^{I_k}\beta\iota_2$, $\nu_k = \beta e_k = \beta P^{I_k}\beta\iota_2$.

Define a sequence of integers $\{k_i\}$, $i \geq 0$, by

$$k_0 = 1,\ k_{2a-1} = p^a - k_{2a-2},\ k_{2a} = k_{2a-1} + a,\quad a \geq 1.$$

This sequence begins 1, p-1, p, p^2-p, p^2-p+2,..., and satisfies

$$a+1 < k_{2a} \leq p^a \quad \text{and} \quad k_{2a-2} < k_{2a-1}.$$

In fact $k_1 = p-1$, $k_2 = p \leq p^1$, and for $a \geq 1$

$$k_{2(a+1)} = k_{2(a+1)-1} + (a+1) = p^{a+1} - (k_{2a} - (a+1)) < p^{a+1};$$

$$k_{2(a+1)} = k_{2(a+1)-1} + (a+1) = (p^{a+1} - k_{2a}) + a+1 \geq a+2; \text{ and}$$

$$k_{2(a+1)-1} = p^{a+1} - k_{2a} \geq p^{a+1} - p^a \geq k_{2a}.$$

Hence $\{k_i\}$ is a strictly increasing sequence. Our interest in $\{k_i\}$ derives from the fact that the associated sequence $\{2k_i(p-1)+1\}$ occurs as the degrees of the nontrivial differentials of the AHss $H^*(X_2,Z/p) \Rightarrow \underleftarrow{K}^*(X_2;Z/p)$.

THEOREM 2.1. Let $e_{11} = x^{p-1}e_1$, $e_{21} = e_2 + (\beta x)\nu_1^{p-1}$, and for $j \geq 2$, $e_{1j} = (x^{p^{j-1}})^{p-1}e_{2,j-1}$, $e_{2j} = e_{1,j-1}\nu_j^{p-1}$. Then in the AHss $H^*(X_2;Z/p) \Longrightarrow \underleftarrow{K}^*(X_2;Z/p)$

(i) $d_{2k_0(p-1)+1} = d_{2(p-1)+1}$ sends $x \to e_1$, $\beta x \to -\nu_1$ and $e_j \to \nu_{j-1}^p$, $j \geq 2$. Hence $E_{2(p-1)+2} \cong E(e_{11},e_{21}) \otimes P(x^p,\nu_j)/\nu_j^p$, $j \geq 2$.

(ii) For $j \geq 2$, $E_{2k_{2j-2}(p-1)+2} \cong E_{2k_{2j-1}(p-1)+1} \cong$

$E(e_{1j},e_{2j}) \otimes P(x^{p^j},\nu_\ell)/\nu_\ell^p$, $\ell \geq j+1$; and

$$(2.1)_j \qquad \begin{cases} d_{2k_{2j-1}(p-1)+1}e_{1j} = \nu_{j+1} \\ d_{2k_{2j}(p-1)+1}x^{p^j} = e_{2j} \end{cases} \qquad j \geq 1.$$

We shall also require the following two results on the associated $Z_{(p)}$ AHss.

THEOREM 2.2. In the $Z_{(p)}$ AHss $H^*(X_2;Z_{(p)}) \Rightarrow K^*(X_2;Z_{(p)})$

$$(2.2)_j \qquad d_{2p^j(p-1)+1}(p^j x^{p^j-1}u_0) = v_{j+1}, \quad j \geq 0.$$

PROPOSITION 2.3. In the $Z_{(p)}$ AHss of 2.2 we have for all $k \geq 1$

$$(2.3)_j \qquad d_r(p^j x^{kp^j-1}u_0) \neq 0 \text{ for some } r \leq 2p^j(p-1)+1, \quad j \geq 0.$$

Proofs of 2.1, 2.2 and 2.3. Except for the initial differential $d_{2(p-1)+1}$, when we write $d_r a = b$ we mean $d_r a = Nb$ for some N with $(N,p) = 1$. In all cases b will be an element of order p, so our statement $d_r a = b$ can be read "a kills the Z/p summand generated by b."

The initial differential of the Z/p AHss is $d_{2(p-1)+1} = P^1\beta - \beta P^1$ and a direct calculation produces the description of E_{2p} given in 2.1(i). Also the initial differential of the $Z_{(p)}$ AHss is $d_{2(p-1)+1} = -\beta_{(p)}P^1$ and a similar direct calculation shows that

$$d_{2(p-1)+1}x^k u_0 = -x^k v_1.$$

This establishes $(2.2)_0$ and $(2.3)_0$.

We now prove the triple of assertions $(2.1)_j$, $(2.2)_j$, $(2.3)_j$ by induction on j. Since the same arguments are used in the proof of the initial step (i.e., $(2.1)_1$, $(2.2)_1$, $(2.3)_1$) as in the general step, we shall prove them simultaneously. Thus we assume the truth of the assertions $(2.1)_i$, $(2.2)_i$, $(2.3)_i$ for all $1 \leq i < j$ when $j > 1$ and we make no assumption when $j = 1$.

From [3] we know that the degree r of any nontrivial differential of either the Z/p or $Z_{(p)}$ AHss converging to inverse limit complex K-theory must be of the form $r = 2s(p-1)+1$, i.e., $r \equiv 1 \bmod 2(p-1)$. Secondly, in [8] it is shwon that the classes $v_j \in H^{2p^j+2}(K(Z,3);Z_{(p)})$ and their mod p reductions are permanent cycles in the $Z_{(p)}$ AHss and the Z/p AHss, respectively. Since the v_j appearing in assertions $(2.1)_j$, $(2.2)_j$, $(2.3)_j$ are π^*-images of their namesakes in $H^*(K(Z,3);G)$, $G = Z/p$ or $Z_{(p)}$, they also are permanent cycles. By acyclicity [1], [4], these v_j are images under some set of higher order differentials.

Under the homeomorphism of M_2 given in ¶1 we consider the skeleta of M_2 defined by

$$M_{2,j} \cong \Sigma\, CP^{p^j} \cup C((CP^\infty * CP^\infty)^{(2p^j+1)}), \quad j \geq 0$$

where $A^{(k)}$ is the k-skeleton of A. Note $M_{2,0} \cong S^3$. An easy calculation in [8] gave the result

$$(2.4)\quad H^i(M_{2,j};Z_{(p)}) \cong \begin{cases} Z_{(p)} & i = 0,3 \\ Z_{(p)}{}^{p^k-2} + Z/p & 3 < i = 2p^k+2,\ k \le j \\ Z_{(p)}{}^{\ell-1} & 4 < i = 2\ell+2 \neq 2p^k+2,\ i \le 2p^j+2 \\ 0 & \text{otherwise} \end{cases}$$

Let Su denote a 3-dimensional generator. The torsion summands have as generators $i^*\nu_k = i^*\beta_{(p)}P^{I_k}u_0 = \beta_{(p)}P^{I_k}Su$, $i:M_{2,j} \to K(Z,3)$. The classes $i^*u_k = i^*P^{I_k}u_0 = P^{I_k}Su$ are nonzero mod p classes and they satisfy $\beta_{(p)}i^*u_k = i^*v_k$.

Just as M_2 was used to study $K(Z,3)$ in [8], here we use $\pi^{-1}M_2$ to study X_2. In particular, the classes $Nx^ku_0 \in H^*(X_2;Z_{(p)})$ are never boundaries in the $Z_{(p)}$ AHss for X_2 (use naturality with respect to $\pi^{-1}M_{2,0} \to X_2$). Hence acyclicity implies that for any class x^ku_0 with $k+1 = p^sk_1$, $(k_1,p) = 1$ (so x^ku_0 has order p^{s+1}), there exist $s+1$ integers $r_1,r_2,\ldots,r_{s+1}$ such that $d_{r_i}(p^{i-1}x^ku_0) \neq 0$ (because the torsion part of $H^*(X_2;Z_{(p)})$ not generated by the classes x^ku_0 is a Z/p vector space and so $d_{r_i}(p^{i-1}x^ku_0)$ is an element of order p for each r_i). Although the $Z_{(p)}$ AHss for $\pi^{-1}M_2$ will not be acyclic, our proof will show that the classes x^kS_u are never boundaries and will have $s+1$ nontrivial differentials defined on the classes p^ix^kSu, so that no multiple of x^kSu survives to E_∞.

Consider $(2.2)_1$. The classes Nx^ku_0 surviving the initial differential $d_{2(p-1)+1}$ are all px^ku_0, $k+1$ a multiple of p. In dimensions less than $2p^2+2-2p+1$ the only odd dimensional classes at E_{2p} are $px^{kp-1}u_0$, $1 \le k \le p-1$. As $\dim v_2 = 2p^2+2 \equiv 4 \bmod 2(p-1)$, $\dim px^{kp-1}u_0 = 2kp+1 \equiv 2k+1 \bmod 2(p-1)$, and for $1 \le k \le p-1$, $2k+1 \equiv 3 \bmod 2(p-1)$ if and only if $k = 1$, then the only possible way for v_2, to die is

$$d_{2p(p-1)+1}px^{p-1}u_0 = v_2.$$

This is $(2.2)_1$. The first half of $(2.2)_j$ follows similarly. Indeed, by induction we have $E_{2k_{2j}(p-1)+2} \cong E(e_{1j},e_{2j}) \otimes P(x^{p^j},v_\ell)/v_\ell^p$, $\ell \ge j+1$. Since

$$\dim e_{2j} = \dim v_j^p - (2k_{2j-3}(p-1)+1) = (2p^j+2)p - 2k_{2j-3}(p-1)-1$$

and

$$\dim v_{j+1} - \dim e_{2j} = 2k_{2j-3}(p-1)+1-2p < 2k_{2j-2}(p-1)+1,$$

then e_{2j} cannot kill v_{j+1} and we must have $d_re_{1j} = v_{j+1}$ for some r. In fact

$$r = 2p^{j+1}+2 - \dim e_{1j} = 2p^{j+1}+2 - (\dim(x^{p^{j-1}})^p + 2k_{2j-2}(p-1)+1)$$

$$= 2p^j(p-1) - 2k_{2j-2}(p-1)+1 = 2k_{2j-1}(p-1)+1,$$

verifying our opening description of the sequence $\{k_i\}$.

For the second half of $(2.1)_j$ we must digress to consider $(2.2)_j$. Our induction hypotheses imply that the only classes $Nx^k u_0$ of dimension less than $2p^{j+1}+2 - 2p^{j-1}(p-1)-1$ surviving to $E_{2p^j(p-1)+1}$ are $p^j x^{kp^{j-1}} u_0$, $1 \leq k \leq p-1$. Of these, only $p^j x^{p^{j-1}} u_0$ can kill v_{j+1} since $2p^{j+1}+2 \equiv 4 \bmod 2(p-1)$ and $2kp^j+1 \equiv 2k+1 \bmod 2(p-1)$, so $k = 1$. But naturality with respect to $\pi^{-1}M_{2,j+1} \to X_2$ shows that elements not of the form $Nx^k u_0$ cannot kill v_{j+1}. Hence we obtain $(2.2)_j$

$$d_{2p^j(p-1)+1}\, p^j x^{p^{j-1}} u_0 = v_{j+1}.$$

Returning to the second half of $(2.1)_j$ we have $\beta_{(p)} x^{p^j} = p^j x^{p^{j-1}} u_0$ and by $(2.2)_j$ $d_{2p^j(p-1)+1}\, p^j x^{p^{j-1}} u_0 \neq 0$. Hence $d_r x^{p^j} \neq 0$ for some $r \leq 2p^j(p-1)+1$. From our induction assumptions and the first half of $(2.1)_j$ we must have $d_{2k_{2j}(p-1)+1} x^{p^j} = e_{2j}$. This completes the proof of $(2.1)_j$ and $(2.2)_j$.

More essential use of the space $\pi^{-1}M_{2,j+1}$ is needed for the proof of $(2.3)_j$. At this stage of the induction we know that for X_2 $(2.3)_i$ for $i < j$ and $(2.1)_i$, $(2.2)_i$ for $i \leq j$ are all true. Suppose then for $\pi^{-1}M_{2,j+1}$ we could show

$$(2.5) \qquad d_{2p^j(p-1)+1}\, p^j x^{kp^{j-1}} Su = x^{(k-1)p^i} i^* v_{j+1}, \quad k \geq 1.$$

Then $(2.3)_j$ would follow immediately from naturality. We establish (2.5) first for $(k,p) = 1$ and then for $(k,p) = p$. By $(2.2)_i$, $i \leq j$ and naturality we have for $\pi^{-1}M_{2,j+1}$ that

$$d_{2p^i(p=1)+1}\, p^i x^{p^{i-1}} Su = v_{i+1}.$$

As $\beta_{(p)} x^{p^i} = p^i x^{p^{i-1}} Su$, this implies for $i \leq j$ that

$$d_{2p^i(p-1)+1}x^{p^i} = i^*u_{i+1}.$$

But then $d_{2p^i(p-1)+1}x^{kp^i} = kx^{(k-1)p^i}i^*u_{i+1}$. Hence for $(k,p) = 1$, $d_{2p^i(p-1)}p^ix^{kp^{i-1}}Su = i^*v_{j+1}$ when $i \leq j$. This gives (2.5) when $(k,p) = 1$.

For the cases $(k,p) = p$ let k_0 be the least integer (necessarily a multiple of p) for which (2.5) fails. Then we claim that the Universal Coefficient Theorem (UCT)

$$0 \to K^0(X;Z_{(p)}) \times Z/p \to K^0(X;Z/p) \to \mathrm{Tor}(K^1(X;Z_{(p)}),Z/p) \to 0$$

fails for the space $X = (M_{2,j+1})^{(2s)}$, $2s = 2p^{j+1}+2 + 2(k_0-1)p^j$. Our induction hypotheses and the assertion that (2.5) fails for k_0 imply that the $Z_{(p)}$ AHss and the Z/p AHss for X are completely determined. In particular, the nontrivial differentials are given as follows:

(i) in the $Z_{(p)}$ AHss

$$\text{(a) } d_{2p^i(p-1)+1}p^ix^{kp^i-1}u_0 = \begin{cases} x^{(k-1)p^i}i^*v_{i+1} & \text{if } i < j \text{ or } k < k_0 \\ 0 & \text{if } i = j \text{ and } k = k_0 \end{cases}$$

(Here and in the following we interpret a class to be zero if its dimension exceeds the dimension of X.);

(ii) in the Z/p AHss

(b) $d_{2(p-1)+1}x = i^*u_1$ and $d_{2(p-1)+1}u_0 = -i^*v_1$

(c) $d_{2p^i(p-1)+1}x^{kp^i} = kx^{(k-1)p^i}i^*u_{i+1}$ $\quad k \geq 1,\ i \geq 1$

(d) $d_{2(p^i-p^{i-1})(p-1)+1}x^{kp^i-p^{i-1}}i^*u_i = x^{(k-1)p^i}i^*v_{i+1}$ $\quad i \geq 1$

As usual we establish the failure of UCT by counting Z/p summands in each of its three terms. From (a) all $x^{\ell}i^*v_{i+1}$ with ℓ not a multiple of p^i survive to E_∞. And from (d) so do their mod p reductions. Reduction mod p shows that no nontrivial group extensions occur among the former $x^{\ell}i^*v_{i+1}$, so the summands generated by these elements appear equally in the first and second terms of the UCT. Secondly there are free generators $x^kw \in H^{even}(X,Z_{(p)})$ where w is a free generator in (2.4). As all differentials are zero on these classes and their mod p reductions, they define an equal number of summands again in the first and second terms of the UCT. Finally there are precisely two additional even dimensional generators whcih survive to E_∞ in this $Z_{(p)}$ AHss namely, $x^{(k_0-1)p^j} i^*v_{j+1}$ and the free generator X^s, $s = p^{j+1} + 1 + (k_0-1)p^j$ (the top dimensional cell coming off the fibre CP^∞ via the truncation). The generator X^s certainly is not involved in a nontrivial group extension since otherwise for $\ell > 2p^{j+1} + 2 + 2(k_0-1)p^j$ the space $(\pi^{-1}M_{1,j+1})^{(\ell)}$ will have all of the above even dimensional classes at E_∞ save X^s and naturality with respect to $(\pi^{-1}M_{2,j+1})^{(2s)} \to (\pi^{-1}M_{2,j+1})^{(\ell)}$ would produce a contradiction. The summand in the first term of the UCT generated by X^s will correspond to the summand of the second term generated by its mod p reduction x^s.

To complete the argument we show that $x^{(k_0-1)p^j} i^*v_{j+1}$ is not involved in any nontrivial group extension and so determines an additional summand in term 1 and that all other survivors to E_∞ define an equal number of summands in terms 2 and 3.

For $K^1(X;Z_{(p)})$ observe that $H^{odd}(X;Z_{(p)})$ is generated by classes of form x^ku_0. Such classes are never boundaries in the $Z_{(p)}$ AHss. We called a class Nx^ku_0 an <u>unstable</u> cycle of a differential d_r if $\dim d_r(Nx^ku_0) > \dim X$ (and so $d_rNx^ku_0 = 0$) and a <u>stable</u> cycle if $d_r(Nx^ku_0) = 0$ while $\dim(Nx^ku_0) \leq \dim X$.

Consider then the classes x^ku_0 in the $Z_{(p)}$ AHss and the classes x^k in the Z/p AHss. Surviving the initial differential $d_{2(p-1)+1}$ in the $Z_{(p)}$ AHss will be a certain number (say N_0) of unstable cycles x^ku_0 and a certain number (say N_0') of stable cycles $px^{\ell p-1}u_0$ (because $d_{2(p-1)+1}x^ku_0$ will be an element of order p in this case). For the same differential in the Z/p AHss there will be N_0+1 unstable cycles x^k (x^s will correspond to the free class X^s and the remaining N_0 x^k will correspond to the N_0 unstable cycles x^ku_0) and N_0' stable cycles $x^{\ell p}$. Clearly the unstable $d_{2(p-1)+1}$ cycles in either spectral sequence survive to E_∞.

The next differential which is nontrivial on classes $px^{\ell p-1}u_0$ in the $Z_{(p)}$ AHss or on classes $x^{\ell p}$ in the Z/p AHss has degree $2p(p-1)+1$ in either case. In the $Z_{(p)}$ case there will be say N_1 unstable cycles $px^{\ell p-1}u_0$ and say N_1' stable cycles $p^2x^{p^2-1}u_0$. And in the Z/p case there will be N_1 unstable cycles $x^{\ell p}$ and N_1' stable cycles $x^{\ell p^2}$. For each succeeding nontrivial differential which is nontrivial on classes of the form

$Nx^k u_0$ in the $Z_{(p)}$ case or in classes x^{kp^i} in the Z/p case the cardinalities of the unstable cycles will be the same as will be the cardinalities of the stable cycles —the reason for this is that at each stage the degrees of the nontrivial differentials in the two spectral sequence are equal.

Hence the additional generators at E_∞ produced this way generate an equal number of summands in the second and third terms of the UCT.

The preceding arguments show that the UCT is satisfied by the summand thus far accounted for. Hence to show that $x^{(k_0-1)p^j} i^*v_{j+1}$ does not survive to E it suffices to show that its presumed survival generates an extra summand in the first term of the UCT. This will be done if we show it is not involved in any nontrivial group extension. Since $\dim x^{(k_0-1)p^j} i^*v_{j+1} = \dim X$, a nontrivial group extension involving $x^{(k_0-1)p^j} i^*v_{j+1}$ can only occur with a class of the form $x^k i^*v_i$, $i \leq j+1$ and k not a multiple of p^{i-1}. But from (b), (c), (d) of (ii) it follows that the classes $x^k i^*u_i$, $i \leq j+1$ and k not a multiple of p^{i-1}, survive to E_∞ and so does the class $x^{(k_0-1)p^j} i^*u_{j+1}$. But $\beta_{(p)}$ takes these classes at the E level of the Z/p AHss onto the classes at the E_∞ level of the $Z_{(p)}$ AHss got by replacing u_i by v_i. Hence a nontrivial group extension cannot occur involving the class $x^{(k_0-1)p^j} i^*v_{j+1}$ and the UCT is not satisfied.

Hence (2.5) is satisfied for all $k \geq 1$ and the proof of $(2.3)_j$ is complete. Observe that $x^{k_0p^j-1}Su$ has order p^t, $t \geq j+2$ since k_0 is a multiple of p. Thus the assertion

$$(2.6) \qquad d_{2p^j(p-1)+1}\, p^j x^{k_0p^j-1} Su = x^{(k_0-1)p^i} i^*v_{j+1}$$

does not reduce the rank of the third term of the UCT from what it would have been had $d_{2p^j(p-1)+1}\, p^j x^{k_0p^j-1} Su = 0$. So with (2.6) we have that UCT is satisfied.

REMARKS. 1. A similar counting argument would prove (2.5) also in the case $(k,p) = 1$.

2. This counting argument could have been avoided either by determining much more of the $Z_{(p)}$ AHss for X_2 and using acyclicity or by studying the $Z_{(p)}$ AHss for a space such as $\pi^{-1}M_{2,j+1} \times \pi^{-1}M_{2,j+1}$ and using

naturality with respect to the multiplication $m\colon X_2 \to X_2 \to X_2$.

3. ELEMENTARY SPECTRAL SEQUENCES

DEFINITION 3.1. A spectral sequence $\{E^{p,q}, d_r\}$ is of Atiyah-Hirzebruch type if $d_r = 0$ for all r not of the form $2k(p-1)+1$. A spectral sequence $\{E_r^{p,q}, d_r\}$ is acyclic if $E_\infty^{p,q} = 0$ for all $(p,q) \neq (0,0)$.

Topological $Z_{(p)}$ or Z/p AHss's are of Atiyah-Hirzebruch type [3]. The results of [1], [4] imply the existence of many acyclic spectral sequences of A-H type. All of these sequences have an initial differential algebra of the form $E \otimes P$, E a Z/p exterior algebra and P a Z/p polynomial algebra.

DEFINITION 3.2. An elementary type I spectral sequence is a spectral sequence of Z/p algebras such that

(i) its differentials are derivations;

(ii) $E_2^{**} \simeq E^{**}_{2(p-1)+1} = E(e_0', e_0'', e_1, e_1, \ldots) \otimes P(z_0', z_0, z_1, z_1, \ldots)$ where $\dim z_i = 2\ell p^i + 2$ for $i \geq 0$, $\dim e_i = \dim z_i - 1$ for $i \geq 1$, and $\dim e_0'' = 2(p+\ell)+1$ (the dimensions of e_0', z_0' will not be signicant).

(iii) $d_{2(p-1)+1}$ is given by

$$\begin{array}{ccccc} e_0' & e_1 & e_1 & e_3 & \\ \downarrow & \vdots & \vdots & \vdots & \\ z_0' & z_0 & z_1 & z_1 & \cdots \\ \downarrow & \downarrow & \downarrow & \downarrow & \\ 0 & e_0'' & 0 & 0 & \end{array}$$

and so $E^{**}_{4(p-1)+1} = E(f_1) \otimes P(z\ )/z\ ^{p}$, $\ell \geq 2$, $f_1 = z_0^{p-1} e_0''$ (here $x \dashrightarrow y$ and $x' \longrightarrow y'$ mean the differential sends x onto y^p and x' onto y', respectively);

(iv) for each $k \geq 2$ $E^{**}_{2k(p-1)+1} \cong E(f_{k-1}) \otimes P(z_\ell)/z_\ell^{p}$, $\ell \geq k$, $f_1 = z_0^{p-1} e_0''$, $f_k = z_k^{p-1} f_{k-1}$ and $d_{2k(p-1)+1}$ is given by

$$\begin{array}{ccc} z_k & z_{k+1} & z_{k+1} \\ \downarrow & \downarrow & \downarrow \\ f_{k-1} & 0 & 0 \end{array}$$

Observe (iv) implies that an elementary type I spectral sequence is acyclic, for as $k \to \infty$ the dimensions of the generators at $E^{**}_{2k(p-1)+1}$ also

$\to \infty$. Let us check the homology of $d_{2(p-1)+1}$ and of $d_{2k(p-1)+1}$, $k \geq 2$. For $d_{2(p-1)+1}$ note that the homologies of the differential algebras $A_0 = E(e_0') \otimes P(z_0')$, $de_0' = z_0'$; $A_1 = E(e_1, e_0'') \times P(z_0)$, $de_1 = z_0^p$, $dz_0 = e_0''$; $A_k = E(e_k) \otimes P(z_{k-1})$, $de_k = z_{k-1}^p$, $k \geq 2$, are 0; $E(z_0^{p-1}e_0^1)$; $P(z_{k-1})/z_{k-1}^p$. So the Künneth formula gives the description of $E^{**}_{2(p-1)+2} \cong E^{**}_{4(p-1)+1}$ given in (iii). Secondly the homology of the differential algebra $A = P(z_k)/z_k^p \otimes E(f_{k-1})$, $d_k = f_{k-1}$ is $E(f_k)$. Again the Künneth formula gives the description of $E^{**}_{2k(p-1)+2} \cong E^{**}_{2(k+1)(p-1)+1}$ given in (iv).

Secondly let us verify that the degree of $d_{2k(p-1)+1}$ is in fact $2k(p-1)+1$, i.e., $\dim f_{k-1} - \dim z_k = 2k(p-1)+1$. Since $f_{k-1} = d_{2k(p-1)+1}z_k$, $\dim z_k^p = pZ_k$ and $\dim f_k = F_k$ satisfy

$$F_k = pZ_k + 2k(p-1)+1.$$

$Z_k = 2(p+\ell-1)p^{k-1}+2$ and so $pZ_k - Z_{k+1} = 2p-2$, whence

$$F_k - Z_{k+1} = pZ_k + 2k(p-1)+1 - Z_{k+1} = 2(k+1)(p-1)+1$$

completing the inductive verification of the degrees of the differentials.

Observe that the degrees of the differentials are independent of ℓ (the generator z_0 has dimension $2\ell+2$).

Before we can indicate when and how elementary type I spectral sequences arise, we need to introduce elementary type II spectral sequences.

DEFINITION 3.2. An <u>elementary type</u> II spectral sequence is a spectral sequence of Z/p algebras such that

(i) its differentials are derivations;

(ii) $E_2^{**} - E^{**}_{2(p-1)+1} = E(e_{01}, e_{02}, e_2, e_3, \ldots) \otimes P(x, z_1, z_2, \ldots)$ where $\dim x = 2\ell$, $\dim e_{01} = 2\ell+1$, $\dim e_{02} = 2(p+\ell-1)+1$, $\dim z_i = 2(p+\ell-1)p^{i-1}+2$ and $\dim e_i = \dim z_i - 1$;

(iii) $d_{2(p-1)+1}$ is given by

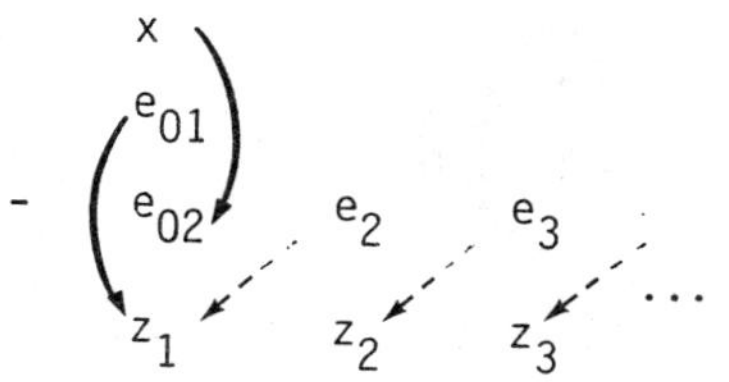

and so $E^{**}_{2(p-1)+2} \simeq E^{**}_{4(p-1)+1} \cong E(e_{11},e_{11}) \otimes P(x^p,z_\ell)/z_\ell^p$, $\ell \geq 2$, where $e_{11} = x^{p-1}e_{02}$, $e_{21} = e_2 + e_{01}z_1^p$;

(iv) for each $a \geq 1$, $E_{2k_{2a-1}(p-1)+1} \cong E(e_{1a},e_{2a}) \otimes P(x^{p^a},z_\ell)/z_\ell^p$, $\ell \geq a+1$, and $d_{2k_{2a-1}(p-1)+1}e_{1a} = z_{a+1}$, $d_{2k_{2a}(p-1)+1}x^{p^a} = e_{2a}$, where $a_{1a} = x^{p^{a-1}(p-1)}e_{2,a-1}$, $e_{2a} = e_{1,a-1}z_a^p$.

As before, we note that (iv) implies acyclicity, since the dimensions of the generators at successive E-levels are increasing without bound. To check that the E-levels are correctly described for the given differentials we use the Kunneth formula and the easy observations that the homology of each of the differential algebras

(i) $A_1 = P(x) \otimes E(e_{02})$, $d(x) = e_{02}$

(ii) $A_2 = E(e_{01},e_2) \otimes P(z_1)$, $d(e_{01}) = -z_1$, $d(e_2) = z_1^p$

(iii) $A_k = E(e_k) \otimes P(z_{k-1})$, $d(e_k) = z_{k-1}^p$

(iv) $B_{1a} = E(e_{1a}) \otimes P(z_{a+1})/z_{a+1}^p$, $d(e_{1a}) = z_{a+1}$

(v) $B_{2a} = P(x^{p^a}) \otimes E(e_{2a})$, $dx^{p^a} = e_{2a}$

is, respectively,

(i)' $P(x^p) \otimes E(e')$, $e' = x^{p-1}e_{02}$

(ii)' $E(e'')$, $e'' = e_2 + e_{01}z_1^{p-1}$

(iii)' $P(z_{k-1})/z_{k-1}^p$

(iv)' $E(e_{2,a+1})$

(v)' $P(x^{p^{a+1}}) \otimes E(e_{1,a+1})$.

The reader should notice that the degrees of the differentials involve the sequence of integers $\{k_i\}$ defined in ¶2. We verify that the degrees of the differentials satisfy the defining equations of the sequence $\{k_i\}$. Setting $Z_i = \dim z_i$ and $E_{1a} = \dim e_{1a}$, we have

$$2k_{2a-1}(p-1) + E_{1a} = Z_{a+1} = 2(p+\ell-1)p^a+2;$$

$$E_{1a} = 2\ell p^{a-1}(p-1) + E_{2,a-1};$$

so, calling the degrees of the differentials in (iv) $2\tilde{k}_r(p-1)+1$, we have

$$2\tilde{k}_{2a-1}(p-1)+1+2\ell p^{a-1}(p-1) + E_{2,a-1} = 2(p+\ell-1)p^a+2;$$

$$2\tilde{k}_{2a-2}(p-1)+1+2\ell p^{a-1} = E_{2,a-1}$$

Hence

$$2(\tilde{k}_{2a-1}+\tilde{k}_{2a-2})(p-1)+2+2\ell p^a = 2(p+\ell-1)p^a+2;$$

$$2(\tilde{k}_{2a-1}+\tilde{k}_{2a-1})(p-1) = 2(p-1)p^a$$

and so $\tilde{k}_{2a-1}+\tilde{k}_{2a-1}$.

Secondly, from $2\tilde{k}_{2a-1}(p-1)+1+E_{1a} = 2(p+\ell-1)p^a+2$ and $2\tilde{k}_{2a}(p-1)+1 + 2\ell p^a = E_{2a}$ we obtain

$$2(\tilde{k}_{2a}-\tilde{k}_{2a-1})(p-1) = E_{1a}+E_{2a}-2\ell p^a-2(p+\ell-1)p^a-2.$$

So it suffices to show that the right hand side is $2a(p-1)$. Assuming

$$E_{1,a-1}+E_{2,a-1} = 2(a-1)(p-1) + 2\ell p^{a-1} + 2(p+\ell-1)p^{a-1}+2,$$

we have

$$\begin{aligned} E_{1a} + E_{1a} &= (2\ell p^{a-1}(p-1) + E_{2,a-1})+(E_{1,a-1}+(2(p+\ell-1)p^{a-1}+2)(p-1)) \\ &= E_{1,a-1}+E_{2,a-1}+2\ell p^{a-1}(p-1)+2(p+\ell-1)p^{a-1}(p-1)+2(p-1) \\ &= 2(a-1)(p-1) + 2\ell p^a + 2(p+\ell-1)p^a + 2(p-1)+2. \end{aligned}$$

This completes the verification.

Again we see that the degrees of the differentials are independent of ℓ.

An example of an elementary type II spectral sequence is the topological Z/p AHss $H^*(X_2;Z/p) \Rightarrow \underset{\leftarrow}{K}^*(X_2;Z/p)$ given in Theorem 2.1. To see how type II spectral sequences arise, consider the standard path fibration $X_{n-1} \to P \to X_n$. The Cartan theorem describing the cohomology algebra $H^*(X_n;Z/p)$ in terms of admissible sequences can be proved by induction on n via a generalized Borel Theorem [7]. In particular if x is an odd admissible generator of $H^*(X_{n-1};Z/p)$, τx will be an even admissible generator of $H^*(X_n;Z/p)$, and if x is an even admissible generator of $H^*(X_{n-1};Z/p)$, $\tau(x^{p^r})$ and $z(x^{p^r})$, $r \geq 0$, will be two sequences of odd, resp., even admissible generators of $H^*(X_n;Z/p)$. Here τ denotes the cohomology transgression. For $z(x^{p^r})$ see [7]. Now $H^*(X_{n-1};Z/p)$, $n > 2$, has many generators x, y satisfying

$d_{2(p-1)+1}x = y^p$ (so that x is necessarily odd dimensional) and $0 \neq \beta x \neq y$. In $H^*(X_n;Z/p)$ we consider the generators

(3.3)

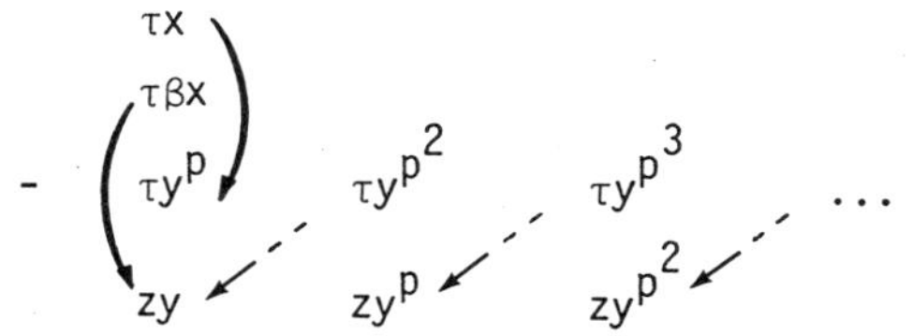

Note we do not include the generators $\tau(\beta x)^{p^r}$, $z(\beta x)^{p^{r-1}}$, $r \geq 1$. Also the behavior of $d_{2(p-1)+1}$ on these generators is indicated. Our conjecture in ¶4 implies that the Z/p AHss $H^*(X_n;Z/p) \Rightarrow \underset{\leftarrow}{K}^*(X_n;Z/p)$ restricted to the subalgebra generated by the classes listed in (3.3) is an elementary type II spectral sequence. For the exceptional fibration $X_1 \to P \to X_2$, we have $x = \iota_1$, $\beta x = y = \beta\iota_1$ and $d_{2p-1}x = y^p$ and the above subalgebra is all of $H^*(X_2\text{-}Z/p)$.

To find a candidate for an elementary type I spectral sequence, apply the above construction to each triple of admissible generates x, βx, y in $H^*(X_2;Z/p)$ with $d_{2(p-1)+1}x = y^p$. Then look at the admissible generators in $H^*(X_3;Z/p)$ not generated from such triples in $H^*(X_1;Z/p)$. These elements are precisely the collection $\{\tau(\iota_2^{p^r}),\ z(\iota_2^{p^r}),\ r \geq 0;\ \tau\beta\iota_2,\ \tau P^1\beta\iota_2,\ \tau\beta P^1\beta\iota_2\}$, i.e., the $\tau(\ )$ and $z(\ )$ elements associated to $\iota, \beta\iota, P^1\beta\iota$ but only the τ-image of $\beta P^1\beta\iota$. It is easy to compute $d_{2(p-1)+1} = P^1\beta - \beta P^1$ on these elements and the result is

(3.4)

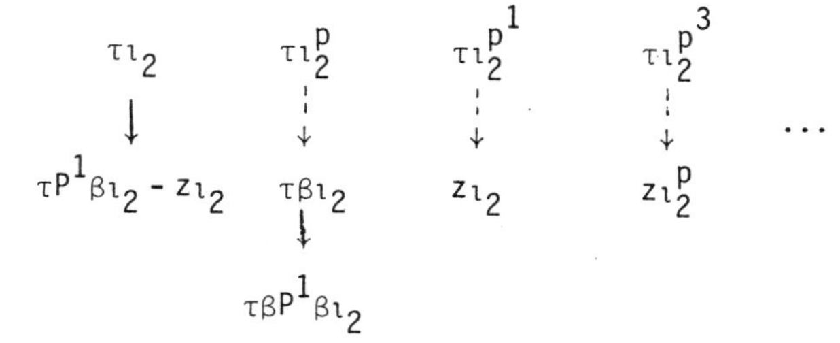

This is the initial differential algebra of an elementary type I spectral sequence. Our conjecture in ¶4 implies that the Z/p AHss $H^*(X_2;Z/p) \Rightarrow \underset{\leftarrow}{K}^*(X_2;Z/p)$ restricted to the subalgebra generated by the elements in (3.4) is an elementary type I spectral sequence. It is not difficult to see from this that the infinite tensor product of this elementary type I spectral sequence and the infinitely many elementary type II spectral sequences generated above (from certain triples in $H^*(X_2:Z/p)$) would give the entire Z/p AHss $H^*(X_3;Z/p) \Rightarrow \underset{\leftarrow}{K}^*(X_3;Z/p)$.

4. CONSTRUCTING THE SPECTRAL SEQUENCE E(n)

For each $n \geq 2$ we construct an algebraic spectral sequence of A-H type whose initial algebra is the cohomology ring $H^*(X_n;Z/p)$. We conjecture that it is the topological AHss $H^*(X_n;Z/p) \Rightarrow \underleftarrow{K}^*(X_n;Z/p)$.

We begin with a triple of elements $e,v,v_1 \in H^*(X_{k-1};Z/p)$ with $v = \beta e$ and $d_{2(p-1)+1}e = v_1^p$. When $k = 2$, $v = v_1$, but otherwise $v \neq v_1$. We refer to such a triple as a dotted arrow $e \dashrightarrow v_1$. To this dotted arrow we associate an infinite family of elementary spectral sequences $E(j,2k+\epsilon;e,v,v_1)$, $1 \leq j \leq k$ and $2k+\epsilon$ runs over all integers ≥ 2 (so $k \geq 1$ and $\epsilon = 0,1$), via the following inductive definition: $E(1,2)(=E(1,2;e,v,v_1))$ is the elementary type II spectral sequence with initial differential algebra

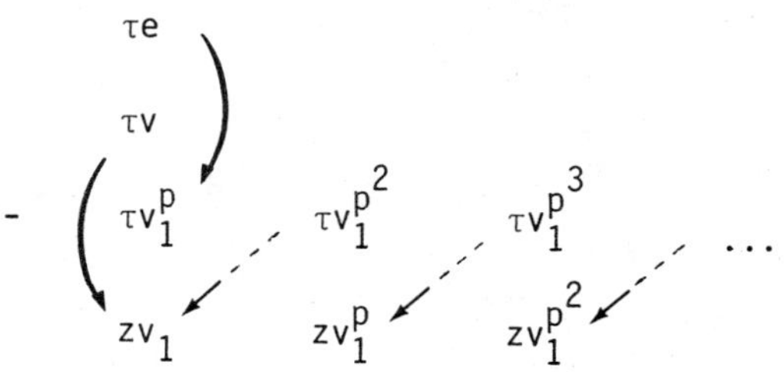

For all $1 \leq j \leq k$ $(k \geq 1)$ assume $E(j,2k)$ has been defined and whose initial differential algebra is given by

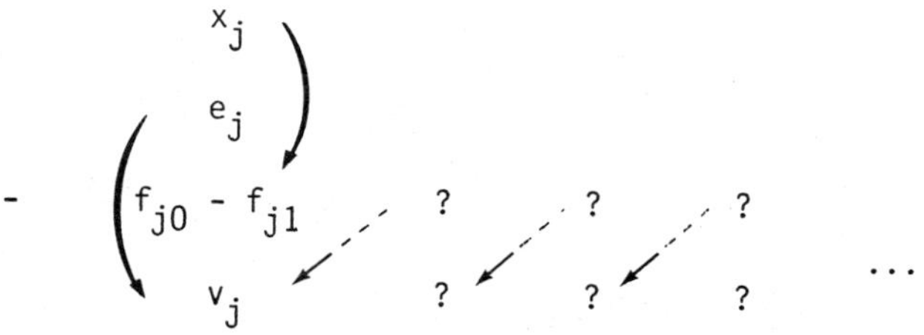

where for $j = k$, $f_{k1} = 0$ and the question marks denote names of elements which do not enter into the construction of $E(j,2k+1)$. Then for $j < k$ $E(j,2k+1)$ is elementary type I spectral sequence with initial algebra

$$
\begin{array}{cccc}
\tau x_j & \tau x_{j+1}^p & \tau x_{j+1}^{p^2} & \tau x_{j+1}^{p^3} \\
\downarrow & \vdots & \vdots & \vdots \\
 & \downarrow & \downarrow & \downarrow \\
\tau f_{j0} - \tau f_{j1} & \tau e_{j+1} & z x_{j+1} & z x_{j+1}^p \\
 & \downarrow & & \\
 & \tau v_{j+1} & &
\end{array}
$$

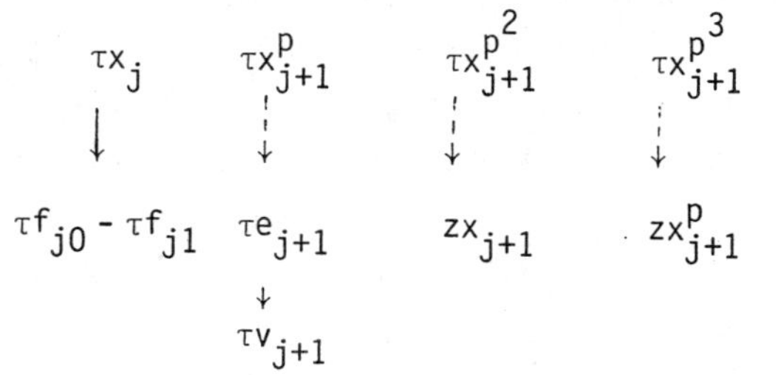

and for $j = k$ the initial algebra of $E(j,2k+1)$ is

$$
\begin{array}{cccc}
\tau x_k & \tau x_1^p & \tau x_1^{p^2} & \tau x_1^{p^3} \\
\downarrow & \vdots & \vdots & \vdots \\
 & \downarrow & \downarrow & \downarrow \\
\tau f_{k0} - zx_1 & \tau e_1 & zx_1 & zx_1^p \qquad \cdots \\
 & \downarrow & & \\
 & \tau v_1 & &
\end{array}
$$

Secondly let us rename these $E(j,2k+1)$ as

$$
\begin{array}{cccc}
e_{j0} & e_{j1} & ? & ? \\
\downarrow & \vdots & \vdots & \vdots \\
 & \downarrow & \downarrow & \downarrow \\
v_{j0} - v_{j1} & v_{j2} & ? & ? \qquad \cdots \\
 & \downarrow & & \\
 & f_j & &
\end{array}
$$

Then for $j \leq k$, $E(j,2k+2)$ is the elementary type II spectral sequence whose initial algebra is

$$
-\begin{pmatrix}
\tau e_{j0} \\
\tau v_{j+1,2} \\
\tau v_{j0} - \tau v_{j1} \\
\tau f_{j+1}
\end{pmatrix}
\begin{array}{ccc}
 & & \\
 & & \\
\tau v_{j0}^p & \tau v_{j0}^{p^2} & \\
zv_{j0} & zv_{j0}^p & \cdots
\end{array}
$$

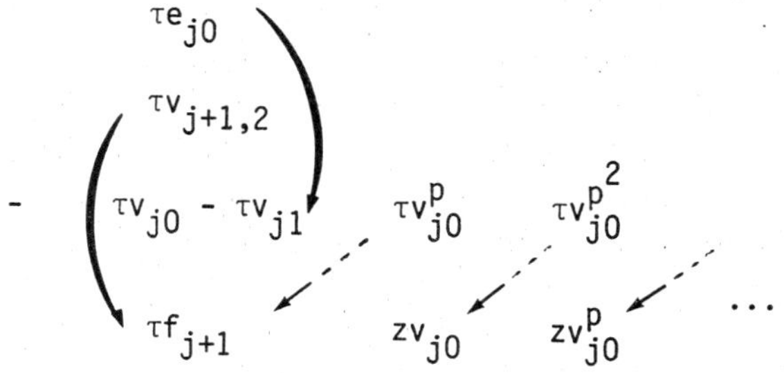

and for $j = k+1$, $E(j,2k+2)$ has initial algebra

$$
-\begin{pmatrix}
\tau e_{k1} \\
\tau v_{k1} \\
\tau v_{k2}^p
\end{pmatrix}
\begin{array}{ccc}
\tau v_{k2}^{p^2} & \tau v_{k2}^{p^3} & \\
zv_{k2}^p & zv_{k2}^{p^2} & \cdots \\
zv_{k2} & &
\end{array}
$$

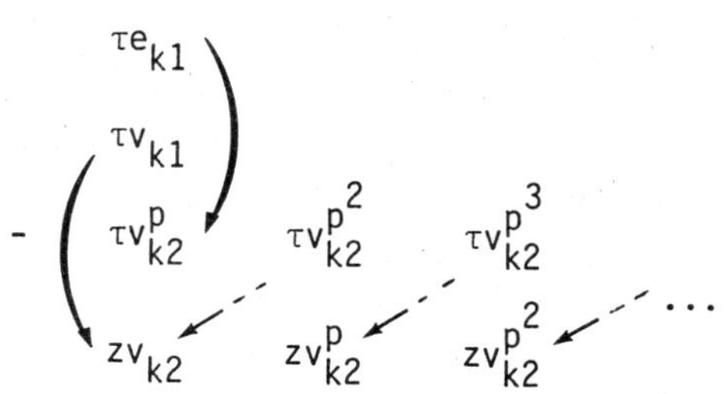

Next we want to iterate the preceding construction. In particular each elementary spectral sequence constructed above has infinitely many dotted arrows and for each dotted arrow we repeat the above construction. And so on.

Warning: This iteration process can and does reproduce some of the previously constructed elementary spectral sequences. For example, the elementary spectral sequences. For example, the elementary type II spectral sequence whose least dimensional element is $P^1\iota_5$ occurs as $E(2,4)$ for the dotted arrow $\iota_1 \dashrightarrow \beta\iota_1$ and as $E(1,2)$ for the dotted arrow $P^1\iota_3 \dashrightarrow \beta\iota_3$, which occurs in $E(1,3)$ for the dotted arrow $\iota_1 \dashrightarrow \beta\iota_1$.

The definition of $E(n)$ will require one further notion. The first stage will produce elementary spectral sequences $E(j_1,2k_1+\varepsilon_1;\iota_1,\beta\iota_1,\beta\iota_1)$ associated to the dotted arrow $\iota_1 \dashrightarrow \beta\iota_1$ in $H^*(X_1;Z/p)$. We abbreviate these to $E(j_1, 2k_1+\varepsilon_1)$. Secondly, to each dotted arrow in any of the $E(j_1,2k_1+\varepsilon_1)$ we construct further elementary spectral sequences $E(j_1,2k_1+\varepsilon_1;j_2,2k_2+\varepsilon_2)$ where our notation has omitted the particular dotted arrow in question. Our definition is as follows: the _degree_ of an $E(j_1,2k_1+\varepsilon_1)$ is $2k_1+\varepsilon_1$ and for $r > 1$ the _degree_ of an $E(j_1,2k_1+\varepsilon_1;j_2,2k_2+\varepsilon_2;\ldots;j_r,2k_r+\varepsilon_r)$ is $\sum_{i=1}^{r}(2k_i+\varepsilon_i)-(r-1)$. $E(n)$ then is defined to be the infinite tensor product of all the distinct spectral sequences in this family of degree = n.

CONJECTURE 1. The Z/p AHss $H^*(X_n;Z/p) \Rightarrow \underleftarrow{K}^*(X_n;Z/p)$ is isomorphic to $E(n)$.

The complexity of these spectral sequences is intimidating. Nonetheless something much more elementary (and perhaps equivalent) can be deduced from this conjecture. In particular there is the notion of p-_rank_ defined for elements of $H^*(X_n;Z/p)$ [6]. For Y_n the Moore space $Y_n = S^n \cup_p e^{n+1}$ there is a filtration of $SP^\infty Y_n \sim X_n$ by the finite symmetric products $\{SP^{p^r}Y_n\}$ and the summand of $H^*(X_n;Z/p)$ consisting of all elements of p-rank $\leq p^r$ is isomorphic to $H^*(SP^{p^r}Y_n;Z/p)$. By comparing the p-ranks of x and y when $d_r x = y$ in the spectral sequence $E(n)$ (i.e., in elementary type I and II sequences), we are led from Conjecture 1 to:

CONJECTURE 2. Filter $SP^\infty Y_n \sim X_n$ by the subspaces $\{Y_n(i)\}_{i \geq -1}$, $Y_n(-1) =$ point, $Y_n(i) = SP^{p^i}Y_n$. Then in the spectral sequence converging to $\underleftarrow{K}^*(X_n;Z/p)$ associated to the filtration $\{Y_n(i)\}$, we have $E_2^{**} \cong E_\infty^{**} = 0$.

From this it would appear that the evaluation of d_1 in the latter spectral sequence corresponds to the evaluation of infinitely many nontrivial differentials in the Z/p AHss.

Finally Conjecture 2 may derive additional interest from the fact that there is a similar collapse of the Rothenberg-Steenrod spectral sequence |1|, |4|. Perhaps Conjecture 2 could be deduced more directly from this result (and then Conjecture 1 from Conjecture 2).

BIBLIOGRAPHY

[1] D.W. Anderson and L. Hodgkin, The K-theory of Eilenberg-MacLane complexes, Topology 7(1968), 317-329.

[2] M.F. Atiyah and F. Hirzebruch, Analytic cycles on complex manifolds, Topology 1(1961), 24-45.

[3] V.M. Buhštaber, Modules of differentials of the Atiyah-Hirzebruch spectral sequence, Math. USSR Sbornik 7(1969), 299-313.

[4] V.M. Buhštaber and A.S. Miščenko, K-theory on the category of infinite cell complexes, Math. USSR Izv. 2(1968), 515-556.

[5] H. Cartan, Algèbres d'Eilenberg-MacLane et homotopie, Séminaire Henri Cartan (1954/1955).

[6] M. Nakaoka, Cohomology mod p of symmetric products of spheres II, J. Inst. Polytech., Osaka Cit- Univ. 10(1959), 67-89.

[7] M.M. Postnikov, On Cartan's Theorem, Russian Surveys 21(1966), 25-37.

[8] J. Ucci, On $BP\langle 1\rangle^*(K(Z,3);Z/p)$, Can. J. Math. 35(1983), 630-653.

Centro de Investigación del IPN
México, D.F.

Syracuse University
Syracuse, New York 13210

Contemporary Mathematics
Volume 58, Part II, 1987

RECENT WORK ON THE PARALLELIZABILITY OF FLAG MANIFOLDS.

P. Zvengrowski[1]

ABSTRACT. The large family of F-flag manifolds over $F = \mathbb{R},\mathbb{C}$, or $\mathbb{H}$ is considered, as well as the related flag^+ manifolds when $F = \mathbb{R}$. The questions of which of these are parallelizable or stably parallelizable are completely solved, apart from a few cases. Techniques used by various authors in obtaining these results are outlined and in some cases illustrated.

1. INTRODUCTION. The solutions to the parallelizability of spheres, projective spaces, and Stiefel manifolds were obtained in the period 1958-1964 by Kervaire [6], Milnor [10], and Sutherland [14]. Interest in the larger class of flag manifolds seems to have started around 1975 with the work of Lam [8] and Yoshida [17]. We will attempt to survey recent developments in this area, based mainly on the Ph.D. thesis of P. Sankaran [11], and also covering results of Miatello-Miatello [9], Trew-Zvengrowski [16], J. Korbaš [1], [7], and R. Stong [13].

The remainder of this section is devoted to defining the flag and flag^+ manifolds as well as clarifying the notation. In §2 a useful tool for obtaining positive results on stable parallelizability or parallelizability, the "λ^2" method, is illustrated. Non-parallelizability results are discussed in §3, and the present knowledge in this area is summarized by Theorems 3.1 - 3.6. The few still unsolved cases are indicated in 3.8.

As usual F denotes the real numbers $\mathbb{R}$, complex numbers $\mathbb{C}$, or quaternions $\mathbb{H}$, and $d = \dim_{\mathbb{R}}(F) = 1,2,4$ respectively. The vector space F^n is equipped with the usual Hermitian inner product. Let $n = n_1 + \cdots + n_s$, $n_i > 0$. A flag of type $n_1,\ldots,n_s$ is a family of s mutually orthogonal subspaces $\sigma_1,\ldots,\sigma_s \subset F^n$ with $\dim(\sigma_i) = n_i$, and $G_F(n_1,\ldots,n_s)$ is the space of all such flags. It has

[1] AMS subject classifications (1980) 55R50, 57R25, 57T15.

the structure of a differentiable manifold of dimension $d/2\,(n^2 - \Sigma\, n_i^2)$ (for details cf. [8]). In case $F = \mathbb{R}$, we often write simply $G(n_1,\ldots,n_r)$. A familiar example is $G_F(n_1,n_2) \cong G_{n_1}(F^n) \cong G_{n_2}(F^n)$, the Grassmann manifold of n_1(or n_2) planes in F^n. We always use "s" for the "length" of the flag and "n" for the dimension of the underlying F-vector space.

Again when $F = \mathbb{R}$ one can consider the "flag$^+$" or "partially oriented" manifolds $G(\tilde{n}_1,\ldots,\tilde{n}_r, n_{r+1},\ldots,n_s)$ in which the first r subspaces $\sigma_1,\ldots,\sigma_r$ are also oriented. There is thus an obvious 2^r-fold covering map

$$G(\tilde{n}_1,\ldots,\tilde{n}_r,n_{r+1},\ldots,n_s) \longrightarrow G(n_1,\ldots,n_s),$$

and both spaces have the same dimension. We prefer the term "flag$^+$", which denotes flag manifolds with additional structure, rather than the possibly misleading term "partially oriented". In case r = s we make the additional requirement that the orientations on $\sigma_1,\ldots,\sigma_s$ induce the standard orientation on $\mathbb{R}^n$. With this convention $G(\tilde{n}_1,\ldots,\tilde{n}_{s-1},n_s) = G(\tilde{n}_1,\ldots,\tilde{n}_s)$, and we often write simply $\tilde{G}(n_1,\ldots,n_s)$ for this case.

The families of flag and flag$^+$ manifolds include many familiar examples such as the Grassmann manifolds, projective spaces, oriented Grassmann manifolds, spheres, Stiefel manifolds, some projective Stiefel manifolds, and the "classical" flag manifolds $G(1,1,\ldots,1)$. They seem to have first been considered in full generality by Lam [8], who also introduced the canonical F-vector bundles $\xi_1,\ldots,\xi_s$ over $G_F(n_1,\ldots,n_s)$. The fibre of ξ_i at the point $(\sigma_1,\ldots,\sigma_s) \in G_F(n_1,\ldots,n_s)$ is simply the n_i-dimensional F-vector space σ_i. Note that $\xi_1 \oplus \cdots \oplus \xi_s \approx n\varepsilon^F$, the n-dimensional trivial F-bundle, and also recall the main result of [8] on the tangent bundle

$$\tau G_F(n_1,\ldots,n_s) \approx \rho \sum_{i<j} \bar{\xi}_i \otimes \xi_j \qquad (\rho = \text{realification}).$$

Finally, note that the bundles $\xi_1,\ldots,\xi_s$ are defined similarly over $G(\tilde{n}_1,\ldots,\tilde{n}_r, n_{r+1},\ldots,n_s)$ and that $\xi_1,\ldots,\xi_r$ are oriented bundles in this case.

2. **EXAMPLES OF POSITIVE RESULTS.** The so called λ^2 (second exterior power) construction seems to be a powerful, yet elementary, tool that unifies the proof of parallelizability or stable parallelizability for many real flag manifolds. Before giving examples of its use, let us recall the bundle isomorphisms

(a) $\lambda^n(\xi \oplus \eta) \approx \sum_{i+j=n} \lambda^i\xi \otimes \lambda^j\eta$,

(b) $\lambda^0\xi \approx \varepsilon$, $\lambda^1\xi \approx \xi$, $\lambda^i\xi = 0$ if $i > \dim \xi$,

(c) for an oriented m-plane bundle ξ, $\lambda^i\xi \approx \lambda^{m-i}\xi$ (by use of the Hodge-star operator).

2.1 Example: $\tilde{G}_3(\mathbb{R}^6)$, the Grassmann manifold of oriented 3-planes in $\mathbb{R}^6$, is stably parallelizable (compare [9], Lemma 3.1).

Proof: Of course $\tilde{G}_3(\mathbb{R}^6) \cong \tilde{G}(3,3)$, so from §1 its tangent bundle $\tau \approx \xi_1 \otimes \xi_2$, where ξ_1, ξ_2 are oriented 3-plane bundles with $6\varepsilon \approx \xi_1 \oplus \xi_2$. Then

$$\begin{aligned} 15\varepsilon \approx \lambda^2(6\varepsilon) &\approx \lambda^2(\xi_1 \oplus \xi_2) \\ &\approx \lambda^2\xi_1 \oplus \lambda^2\xi_2 \oplus (\lambda^1\xi_1 \otimes \lambda^1(\xi_2)) \text{ by (a), (b) above,} \\ &\approx \lambda^1\xi_1 \oplus \lambda^1\xi_2 \oplus (\xi_1 \otimes \xi_2) \quad \text{by (c), (b) above} \\ &\approx \xi_1 \oplus \xi_2 \oplus \tau \approx 6\varepsilon \oplus \tau. \end{aligned}$$

2.2 Corollary: $\tilde{G}_3(\mathbb{R}^6)$ is parallelizable.

Proof: Since this space is a π-manifold (i.e. stably parallelizable) and has dimension 9, it follows from the Bredon-Kosinski theorem [2] that either span $\tilde{G}_3(\mathbb{R}^6)$ = span$(S^9) = 1$ or $\tilde{G}_3(\mathbb{R}^6)$ is parallelizable. The double covering $\tilde{G}_3(\mathbb{R}^6) \longrightarrow G_3(\mathbb{R}^6)$ being a local diffeomorphism, span $\tilde{G}_3(\mathbb{R}^6) \geq$ span $G_3(\mathbb{R}^6)$. But span $G_3(\mathbb{R}^6)$ can be shown to be at least 2 by computing $w_8 = 0$ and $k(G_3(\mathbb{R}^6)) = 0$ (cf. [15], p.650). Hence span $\tilde{G}_3(\mathbb{R}^6) \geq 2$ and it must be parallelizable.

2.3 Example: The classical flag manifolds $G(1,\dots,1)$ are parallelizable.

Proof: Here $\tau \approx \sum_{i<j} \xi_i \otimes \xi_j$, where ξ_i are line bundles with

$\xi_1 \oplus \cdots \oplus \xi_n \approx n\varepsilon$. Noting that $\lambda^2\xi_i = 0$, we then have

$$\begin{bmatrix} n \\ 2 \end{bmatrix}\varepsilon \approx \lambda^2(n\varepsilon) \approx \lambda^2(\xi_1 \oplus \cdots \oplus \xi_n) \approx \sum_{i<j} \lambda^1\xi_i \otimes \lambda^1\xi_j \approx \tau.$$

2.4 <u>Example</u>: $G(\tilde{2},1,\ldots,1)$ is parallelizable, $s \geq 3$.

Proof: For brevity write $M = G(\tilde{2},1,\ldots,1)$. Here $n = s+1 \geq 4$, and $m = \dim M = \frac{1}{2}(n^2-4-(s-1)) = \begin{bmatrix} n \\ 2 \end{bmatrix} - 1$. Define a map

$$g : M \longrightarrow G(\tilde{1},\tilde{m}) \cong S^{\binom{n}{2}-1} = S^m$$

by $g(A) = (\Lambda^2\sigma_1, U_A)$, where $A = (\sigma_1,\ldots,\sigma_s) \in M$ and U_A is the orthogonal complement of $\Lambda^2\sigma_1$ in $\Lambda^2\mathbb{R}^n = \mathbb{R}^{\binom{n}{2}}$. Note that $\Lambda^2\sigma_1$ is oriented by $a_1 \wedge a_2$, where $\{a_1,a_2\}$ is any positively oriented basis of σ_1, hence U_A also has a natural orientation.

Let us denote the canonical bundles over $G(\tilde{1},\tilde{m})$ by η_1,η_2. Thus η_1 is the normal bundle of $S^m \subset \mathbb{R}^{m+1}$ and η_2 the tangent bundle. In fact g "preserves" the tangent bundles, more precisely $g^*(\tau S^m) = g^*(\eta_2) \approx \tau M$, for g can be covered by a bundle map

$$\hat{g} : TM \longrightarrow E\eta_2$$

where $\hat{g}(A, a_i \otimes a_j) = (g(A), a_i \wedge a_j)$. Here $i < j$, $a_i \otimes a_j \in \sigma_i \otimes \sigma_j \subset (TM)_A$, and clearly $a_i \wedge a_j$ is orthogonal to $\Lambda^2\sigma_1$, hence lies in $U_A = (E\eta_2)_{g(A)}$.

To complete the proof it will therefore suffice to show $g \simeq 0$ (cf. [5], Ch.III, Theorem 4.7). Since g depends only on σ_1, it clearly factors as $g = g_1\pi$:

$$\begin{array}{ccc} M = G(\tilde{2},1,\ldots,1) & \xrightarrow{\;g\;} & S^m \\ \downarrow \pi & \nearrow g_1 & \\ \tilde{G}_2(\mathbb{R}^n) = G(\tilde{2},s-1) & & \end{array},$$

where π is the fibre map $\pi(\sigma_1,\sigma_2,\ldots,\sigma_s) = (\sigma_1, \mathrm{span}\{\sigma_2,\ldots,\sigma_s\})$. Now $\dim \tilde{G}_2(\mathbb{R}^n) = 2(n-2) < \begin{bmatrix} n \\ 2 \end{bmatrix} - 1 = m$ whenever $n \geq 4$, so g_1 and hence g are null homotopic.

2.4 Remarks: (i) When $s = 2$, $G(\tilde{2},1) \cong S^2$ is not parallelizable.

(ii) The above technique can be applied to prove parallelizability of many other flag$^+$ manifolds (cf. [11], Theorem 18.1).

(iii) When $s-1$ is even (say $s-1 = m = 2r$), the 2^m-fold covering map $V_{n,m} = G(\tilde{2},\tilde{1},\ldots,\tilde{1}) \longrightarrow G(\tilde{2},1,\ldots,1) = M$ factors through the projective Stiefel manifold $X_{n,m}$ (obtained from $V_{n,m}$ by identifying each m-frame with its negative), since the even parity of m implies that each m-frame has the same orientation as its negative (so the orthogonal complement σ_1 also has well-defined orientation):

$$\begin{array}{ccc} V_{n,m} & \longrightarrow & M \\ & \searrow \quad \nearrow & \\ & X_{n,m} & \end{array}$$

All maps being covering projections, it follows that $X_{n,m} = X_{2r+2,2r}$ is parallelizable.

3. THE MAIN THEOREMS. In this section we outline the parallelizability results for flag and flag$^+$ manifolds. In each case some indication of the methods of proof of the authors who have worked on the problem is given.

3.1 Theorem: $G_F(n_1,n_2)$ (equivalently $G_{n_1}(F^n)$ or $G_{n_2}(F^n)$, where $n = n_1 + n_2$) is stably parallelizable if and only if $n_1 = n_2 = 1$, or in case $F = \mathbb{R}$, $n = 2,4,8$, and $n_1 = 1$ (or $n_2 = 1$). Moreover it is parallelizable only in the latter three cases.

The case $n_1 = 1$ goes back to the work of Milnor [10] and Kervaire [6], since $G_1(F^n) \cong FP^{n-1}$. The complete results for $F = \mathbb{R}$ were first given by Yoshida [17], using a geometrical lemma to reduce the problem to studying $KO(\mathbb{R}P^{n_2})$, and in one case ($G_3(\mathbb{R}^8)$) also employing Stiefel-Whitney classes. Further proofs for the real case, using only Stiefel-Whitney classes, were given by Hiller-Stong [4] and Bartík-Korbaš [1]. In Trew-Zvengrowski [16] a uniform approach for $F = \mathbb{C}$, $\mathbb{H}$ as well as $F = \mathbb{R}$ was given, using the iterated inclusion

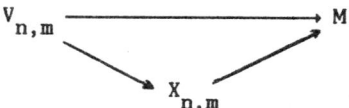

$$FP^{n-k} \cong G_1(F^{n-k+1}) \subset \cdots \subset G_{k-1}(F^{n-1}) \subset G_k(F^n),$$

and the KO-theory of FP^{n-k}.

The flag manifolds with $s > 2$ are covered by the following theorem.

3.2 <u>Theorem</u>: For $s > 2$, $G_F(n_1,\ldots,n_s)$ is stably parallelizable if and only if $n_1 = \cdots = n_s = 1$, and is parallelizable if in addition $F = \mathbb{R}$.

This theorem was proved for $F = \mathbb{R}$ by Korbaš [7] in 1984, using Stiefel-Whitney classes. It was also proved in 1984 by Sankaran [11] (see also [12]) for $F = \mathbb{R},\mathbb{C}$ or $\mathbb{H}$ using the iterated inclusion $G_F(n_1,n_2) \subset G_F(n_1,n_2,n_3) \subset \cdots \subset G_F(n_1,n_2,\ldots,n_s)$ to reduce the problem to the $s = 2$ case, for which 3.1 applies. We have also proved part of this theorem here (cf. 2.3 above).

We now turn to the flag$^+$ manifolds and therefore restrict F to $\mathbb{R}$ in what follows. First consider the $s = 2$ case, omitting the well known case $n_1 = 1$ or $n_2 = 1$ (i.e. the spheres).

3.3 <u>Theorem</u>: Let $n_1,n_2 \neq 1$. $\tilde{G}(n_1,n_2)$ (equivalently $\tilde{G}_{n_1}(\mathbb{R}^n)$ or $\tilde{G}_{n_2}(\mathbb{R}^n)$) is stably parallelizable (or parallelizable) precisely when $G(n_1,n_2)$ is, except that in addition the cases $n_1 = n_2 = 2,3$ are stably parallelizable and $n_1 = n_2 = 3$ parallelizable.

This was proved by Miatello-Miatello [9] in 1982, using Stiefel-Whitney and Pontrjagin classes, Schubert calculus, Lie groups, and in some cases long computations. There may be a gap in their proof at one point. Sankaran [11] obtained a proof of this theorem in 1984 by using a map $g : \mathbb{C}P^2 \longrightarrow \tilde{G}_k(\mathbb{R}^n)$ related to the classifying map for the canonical complex line bundle over $\mathbb{C}P^2$ and then considering $g^* : KO(\tilde{G}_k(\mathbb{R}^n)) \longrightarrow KO(\mathbb{C}P^2)$. He must also resort to calculation of the Stiefel-Whitney classes for the special case $\tilde{G}_4(\mathbb{R}^8)$. Note that the special case $\tilde{G}_2(\mathbb{R}^4) \cong S^2 \times S^2$ (cf. [3], p.104) while $\tilde{G}_3(\mathbb{R}^6)$ was treated in 2.1 above.

We now consider flag$^+$ manifolds with $s > 2$. Without loss of generality we suppose henceforth $n_1 \geq \cdots \geq n_r$, $n_{r+1} \geq \cdots \geq n_s$. The next theorem treats the "fully oriented" case $r = s$ (or equivalently $r = s-1$). The extra assumption $n_2 > 1$ is made to eliminate the well known case of Stiefel manifolds $\tilde{G}(n_1,1,\ldots,1) \cong V_{n,s-1}$, $s-1 \geq 2$, which are parallelizable [14].

3.4 Theorem: Let $s > 2$, $n_2 > 1$. Then $\tilde{G}(n_1,\ldots,n_s)$ is stably parallelizable only if $\{n_1,\ldots,n_s\} \subset \{1,2\}$ or $\{1,3\}$. Moreover $\tilde{G}(2,\ldots,2,\underbrace{1,\ldots,1}_{m})$ is parallelizable if and only if $m \geq 2$, while $\tilde{G}(3,\ldots,3,\underbrace{1,\ldots,1}_{m})$ is parallelizable for all $m \geq 0$.

The first part of this theorem and portions of the second part were obtained in [9]. The results were extended in [11] to the above theorem, except for $\tilde{G}(3,\ldots,3)$ and $\tilde{G}(3,\ldots,3,1)$ whose parallelizability was proved by Stong [13]. Both Miatello–Miatello and Sankaran use the "inclusion" method for the negative results on stable parallelizability. Sankaran uses the λ^2 method for the parallelizability results, and Stong solved the $\tilde{G}(3,\ldots,3)$, $\tilde{G}(3,\ldots,3,1)$ cases (previously known to be π-manifolds) by showing that their Kervaire semicharacteristic vanishes.

3.5 Corollary: Consider $G(\tilde{n}_1,\ldots,\tilde{n}_r,n_{r+1},\ldots,n_s)$. Suppose at least two n_i's are greater than 1, and $\{n_1,\ldots,n_s\} \not\subset \{1,2\}$ or $\{1,3\}$. Then $G(\tilde{n}_1,\ldots,\tilde{n}_r, n_{r+1},\ldots,n_s)$ is not stably parallelizable.

Proof: Its covering space $\tilde{G}(n_1,\ldots,n_s)$ is not stably parallelizable.

Our final theorem treats all the remaining flag$^+$ manifolds, for which some but not all the subspaces are oriented. Again a start towards this theorem was made in [9] and the results greatly extended in [11], the techniques being similar to those in 3.4. By Cor. 3.5. we need only consider cases for which $\{n_1,\ldots,n_s\} \subset \{1,2\}$ or $\{1,3\}$, or precisely one $n_q > 1$.

3.6 Theorem: Let $n_q > 1$ and $n_i = 1$, $i \neq q$. Then $G(\tilde{n}_1,\ldots,\tilde{n}_r,n_{r+1},\ldots,n_s)$ is not stably parallelizable in the following cases:

a(i) $n_{r+1} = 1$, $1 \leq r \leq s-2$, and $n_q \neq 2,6$,

a(ii) $n_{r+1} = 1$, $1 \leq r \leq s-4$, and $n_q \neq 2$,

a(iii) $n_{r+1} > 1$ and $1 \leq r \leq s-3$,

a(iv) $r = s-2$ and $n_q(=n_{s-1}) \neq 3,7$.

a(v) $r = s-2$, $\{n_1,\ldots,n_s\} = \{1,3\}$, and $n_1 = n_{s-1} = 3$.

$G(\tilde{n}_1,\ldots,\tilde{n}_r,n_{r+1},\ldots,n_s)$ is parallelizable in the following cases:

b(i) $G(\tilde{1},\ldots,\tilde{1},1,\ldots,1)$,

b(ii) $G(\underbrace{\tilde{2},\ldots,\tilde{2}}_{k},\underbrace{1,\ldots,1}_{m})$, $m \geq 2$ (see also Remark (ii) below),

b(iii) $G(\tilde{3},\ldots,\tilde{3},\underbrace{\tilde{1},\ldots,\tilde{1}}_{m})$, $m \geq 2$,

b(iv) $G(\tilde{6},1,1)$ and $G(\tilde{6},\tilde{1},1,1)$ (see Remark (ii) below).

3.7 Remarks (i) In (a) above recall that $n_{r+1} = 1$ implies $n_{r+1} = n_{r+2} = \cdots = n_s = 1$ due to our order assumptions on n_i's.

(ii) In b(ii) and b(iv) above extra orientations on the flags can be added at will. The resulting covering spaces will still be parallelizable.

(iii) Theorem 3.6 leaves a few cases unsolved, which we now list.

3.8 Unsolved Cases of Stable Parallelizability:

$G(\tilde{6},\tilde{1},\ldots,\tilde{1},1,1)$, $s \geq 5$,

$G(\tilde{6},\tilde{1},\ldots,\tilde{1},1,1,1)$, $s \geq 4$,

$G(\tilde{1},\tilde{1},\ldots,\tilde{1},a,1)$, $s \geq 3$, $a = 3$ or 7.

REFERENCES

1. V. Bartík and J. Korbaš, "Stiefel-Whitney Characteristic Classes and Parallelizability of Grassmann Manifolds", Proceedings of the 12th Winter School on Abstract Analysis, Supplemento ai Rendiconti del Circolo Matematico di Palermo II 6 (1984), 19-29.

2. G.E. Bredon and A. Kosinski, "Vector Fields on π-Manifolds", Ann. Math. 84 (1966), 85-90.

3. W. Greub, S. Halperin, and R. Vanstone, Connections, Curvature, and Cohomology, Vol. II (1973), Academic Press.

4. H. Hiller and R. Stong, "Immersion Dimension for Real Grassmannians", Math. Ann. 255 (1981), 361-367.

5. D. Husemoller, Fiber Bundles, McGraw-Hill (1966), New York.

6. M. Kervaire, "Non-parallelizability of the n-sphere for n > 7", Proc. Nat. Acad. Sci. U.S.A. 44 (1958), 504-537.

7. J. Korbaš, "Vector Fields on Real Flag Manifolds", preprint (1984).

8. K.Y. Lam, "A Formula for the Tangent Bundle of Flag Manifolds and Related Manifolds", Trans. Amer. Math. Soc. 213 (1975), 305-314.

9. I.D. Miatello and R.J. Miatello, "On Stable Parallelizability of $\tilde{G}_{n,k}$ and Related Manifolds", Math. Ann. 259 (1982), 343-350.

10. J. Milnor, "Some Consequences of a Theorem of Bott", Ann. Math. (2) 68 (1958), 444-449.

11. P. Sankaran, Vector Fields on Flag Manifolds, Thesis, University of Calgary (1985).

12. P. Sankaran and P. Zvengrowski, "On Stable Parallelizability of Flag Manifolds", Pac. J. Math. (to appear).

13. R. Stong, "Semicharacteristics of Oriented Grassmannians", J. Pure Appl. Alg. 33 (1984), 97-103.

14. W. Sutherland, "A Note on the Parallelizability of Sphere Bundles over Spheres", J. Lond. Math. Soc. 39 (1964), 55-62.

15. E. Thomas, "Vector Fields on Manifolds", Bull. Amer. Math. Soc. 75 (1969), 643-683.

16. S. Trew and P. Zvengrowski, "Non-parallelizability of Grassmann Manifolds", Can. Math. Bull. (1) 27 (1984), 127-128.

17. T. Yoshida, "Parallelizability of Grassmann Manifolds", Hiroshima Math. J. 5 (1975), 193-196.

Department of Mathematics and Statistics
The University of Calgary
Calgary, Alberta T2N 1N4
Canada

ABCDEFGHIJ — 8987